Touch in Museums

Touch in Museums

Policy and Practice in Object Handling

Edited by
Helen J. Chatterjee

Book Consultants
Sally MacDonald, David Prytherch, Guy Noble

BERG
Oxford • New York

English edition
First published in 2008 by
Berg
Editorial offices:
First Floor, Angel Court, 81 St Clements Street, Oxford OX4 1AW, UK
175 Fifth Avenue, New York, NY 10010, USA

Berg is the imprint of Oxford International Publishers Ltd.

Library of Congress Cataloging-in-Publication Data

Touch in museums : policy and practice in object handling / edited by Helen J. Chatterjee ; book consultants, Sally MacDonald, David Prytherch, Guy Noble. — English ed.
p. cm.
Includes bibliographical references and index.
ISBN 978-1-84788-239-4 (cloth) — ISBN 978-1-84788-238-7 (pbk.)
1. Museums—Social aspects. 2. Museums—Psychological aspects. 3. Touch—Social aspects. 4. Touch—Psychological aspects. 5. Museum techniques. 6. Museums—Management. I. Chatterjee, Helen. II. MacDonald, Sally. III. Prytherch, David. IV. Noble, Guy.

AM7.T68 2008
069'.068—dc22

2008031218

British Library Cataloguing-in-Publication Data

A catalogue record for this book is available from the British Library.

ISBN 978 1 84788 239 4 (Cloth)
978 1 84788 238 7 (Paper)

Typeset by JS Typesetting Ltd, Porthcawl, Mid Glamorgan
Printed in the United Kingdom by Biddles Ltd, King's Lynn

www.bergpublishers.com

Contents

Illustrations and Tables

Tables

Figures

Contributors

Bernie Arigho, Age Exchange, The Reminiscence Centre, London SE3 9LA.
Michael Boyd, Department of Archaeology, University of Sheffield.
Fiona Candlin, Birkbeck, University of London and British Museum.
Helen Chatterjee, UCL Museums & Collections and Department of Biology, University College London.
Hugo Critchley, Brighton and Sussex Medical Schools, University of Sussex.
Roger Doonan, Department of Archaeology, University of Sheffield.
Alberto Gallace, Department of Psychology, University of Milan-Bicocca, Italy.
Christos Giachritsis, Behavioural Brain Sciences Centre, School of Psychology, University of Birmingham.
Janis Jefferies, Goldsmiths Digital Studios, Goldsmiths, University of London.
Celine Loscos, University of Girona, Spain.
Bernadette Lynch, Writer and Consultant
Francis McGlone, Department of Neurological Science, University of Liverpool.
Guy Noble, UCLH Arts, University College London Hospitals NHS Foundation Trust.
Isil Onol, School of Art, Design and Architecture, University of Huddersfield
Jackie O'Sullivan, Voluntary Services, Newham University Hospital NHS Trust.
Laura Phillips, Department of Coins and Medals, British Museum, London.
Devorah Romanek, University College London and British Museum, London.
Michael Rowlands, Department of Anthropology, University College London.
Jane Samuels, Department of Learning and Audiences, British Museum, London.
Caroline Selai, Sobell Department of Motor Neuroscience and Movement Disorders, Institute of Neurology, University College London.
Charles Spence, Crossmodal Research Laboratory, Department of Experimental Psychology, University of Oxford.
Mandayam Srinivasan, MIT Touch Lab, Research Laboratory of Electronics, Department of Mechanical Engineering, MIT, USA.
Marcus Weisen, Consultant and Trainer, Museums and Galleries Without Barriers, London.
Graeme Were, UCL Museums and Collections, University College London.
Robert Zimmer, Goldsmiths Digital Studios, Goldsmiths, University of London.

Introduction

Helen J. Chatterjee

The value of object handling in the heritage context is not well understood despite the weight of anecdotal evidence suggesting the benefits of physical interaction with objects. An object in the heritage sense includes any item such as an artefact, art work, specimen, manuscript, archive or other historical text. Objects are used in museums and galleries for a range of purposes: enjoyment, to stimulate subject specific interest, knowledge acquisition and inspiration. Understanding the value of object handling is therefore of critical interest and importance to museum practitioners since it forms at least part of many museums' access policies.

Previous writings have focused on the social life of touch (Classen 2005), the history of touch (Classen and Howes 2006) and the barriers facing blind and partially sighted museum visitors (Candlin 2003, 2004, 2006). These fascinating contributions have used historical and practice-led research to broaden our understanding of the social context of touch in cultural heritage. Pye (2008), for example, explores how touch can be used in cultural institutions to facilitate understanding and learning. Beyond these texts there is considerable research in the fields of neurology and psychology on the science of touch (e.g. Heller and Schiff 1991).

This book sets out a framework for understanding the role of object handling for learning, enjoyment and health. It represents an extension of some of the themes explored in Pye's volume (2008) and provides new research on, for example, the emotions of touch, innovative touch technologies including haptic devices, and the role of touch in heritage as a social intervention tool. The broad range of essays explores the many different contexts for object handling, not only within the museum but extending beyond it to hospitals, prisons and the wider community. Further, the volume draws on innovative research from the area of psychoneurology to understand the psychology of touch, and the underlying mechanisms behind physical stimulation and its link with emotions experienced during object handling. The combination of theoretical analysis, policy assessment and detailed case material presented here provides a powerful message of the importance of object handling and the potential for museums to play a vital role in knowledge transfer, social care and health care.

The chapters presented in this volume are the result of a series of workshops organized by the editor in 2006–7. Funded by the UK's Arts and Humanities Research Council, the workshops brought together a diverse range of experts from academic and museum environments, with a view to establishing a better understanding of the value of object handling in museums. Scientists shared their knowledge of the underlying psychological and neurological mechanisms behind touch and sensation. Museum staff discussed the types of practical applications of touch employed in museums such as interactive displays, handling boxes and new technologies for interpreting objects. The union of these two groups afforded a unique opportunity to understand the true value of object handling; the result is this volume, which seeks to provide a valuable toolkit for improving access and interpretation.

In the first part of this volume Candlin begins with a historical perspective, reminding us of the occularcentric (visually-orientated) world around us. Of all of the senses, vision has been determined – in the Western part of the world at least – to be the dominant sense (Classen 2005). The impact for museums has been profound with access to collections almost entirely based on visual provision (exhibitions, displays, etc.). Historically, the tactile experience has been preserved for the privileged few (in the case of nineteenth-century museums), and even today physical access to collections is often treated as a special activity rather than a right.

The psychoneurological perspectives detailed by Critchley, McGlone, Spence and Gallace in the first part of the volume outline the key physical and emotional mechanisms involved in touch. The experimental research discussed in these chapters provides evidence to suggest there are physiological aspects of touch directly linked to emotional feelings which guide and motivate behaviour and reinforce memories. This 'emotional touch' is critically important within heritage since it affords strong support for the value of physical interaction with objects, rather than just visual. The scientific evidence also makes clear and strong links between the various senses (vision, touch, smell, hearing, taste), encouraging a multisensory approach to museum access. Particular highlights in this section include comments from a neuroscience perspective as to how curators could better select objects for handling and how museums could guide visitors to use different tactile exploratory behaviours. Part I also outlines key areas of research needed in order for museums to provision multisensory needs; this is particularly important for blind and partially sighted visitors where previous research has shown that current provision is often seriously insufficient.

A significant proportion of this book is dedicated to haptic technologies – that is, the field of tactile exploration involving technological interfaces. Giachritsis discusses the major psychophysical methodologies employed in researching haptic perception, together with a useful overview of selected current technologies, their potential and limitations. Other authors in this section (Doonan and Boyd, Were, Loscos, Zimmer, Jefferies and Srinivasan) all describe project-based examples of newly implemented haptic technologies in specific settings, making useful case studies for reference. Were, Doonan and Boyd provide creative examples of the potential of haptic interventions in higher education and their role in knowledge transfer. Loscos, Zimmer, Jefferies and Srinivasan present research on the development of new digital technologies which seek to enhance heritage interactions. Onol's 'Tactual Exploration', in this section, further exemplifies the importance of multisensory provisioning. She recommends creating a layered approach to display, tactile opportunities and technology which services a broad cross-section of society, from children to adults, culturally distinct visitors, and people with cognitive, visual, physical and hearing impairments. This exemplar project serves as a case study of best practice in object interpretation, display and museum education.

Part III of the book explores the role of tactile memory and object handling in reminiscence work. Gallace and Spence discuss research which has led to a greater understanding of the neurocognitive mechanisms implicit in people's memory for touch. Importantly, an appreciation of the role of memory in touch affords Gallace and Spence an insightful view of the potential for using touch to greater advantage in museums. For example, their research has revealed that longer exploration times are needed when people have to remember details concerning the tactile stimuli they have manipulated. Further, they suggest that the active manipulation of objects is preferable to passive tactile stimulation. Combined, these assertions demonstrate the value of affording an interactive tactile experience which is not time limited.

Experiential approaches to using objects to stimulate memories are provided by Arigho, Rowlands and Phillips, but from very different perspectives. Arigho, as Director of the Age Exchange programme (one of the UK's most successful reminiscence centres), outlines best practice in the use of objects in reminiscence work which, crucially, is person (not object-)led. Rowlands, an anthropologist, examines the role of 'care as curation' in the context of elderly people moving into sheltered accommodation. Phillips describes a curatorial perspective on the use of objects in handling sessions at the British Museum involving older visitors. The project demonstrates the great value afforded by objects in focusing on the remembered past, but also as tools for stimulating new learning and experiences. Phillips provides a valuable word of caution regarding the elicitation of negative memories during object handling and the need to exercise care in dealing with the emotional responses to touch; a warning that is repeated in the subsequent section of this book, dealing with object handling and therapy.

The therapeutic potential of object handling is tackled in Part IV. There is considerable evidence demonstrating the links between the visual arts and patient recovery time, stress, the need for medication and pain intensity. Research has shown that improving the quality of the hospital environment and enriching patients' lives while undergoing treatment has a positive impact on a host of psychological and physiological variables (for a review of this literature see Noble and Chatterjee, this volume). Recent research presented in this section by Noble and Chatterjee provides compelling evidence of the valuable role object handling can play in therapy and enrichment in clinical settings. This preliminary research found that patients in object-handling sessions revealed significantly more information about their psychological state compared to standard counselling sessions. Further, that specific objects evoked deep emotional responses suggesting that some patients engaged with the objects on a symbolic level, perhaps revealing more than they would under normal circumstances. In such instances, it was apparent that the object became the focus for uncovering emotions, acting as vehicle via which the patient felt more able to communicate intimate thoughts. This research echoes the sentiments highlighted by Phillips's work at the British Museum of the need for caution in approaching object handling sensitively. Whilst object interactions in such settings as hospitals and care homes may be highly beneficial, it is vital these dialogues are undertaken within the context of a complete psychological support mechanism such that any confounding or negative effects are accommodated. This may include appropriate training for staff in museums and/or the need for close links with facilities able to provide psychological support. The above provides a strong argument for closer collaborations between museums and social care organizations.

An inspiring case study of good practice in this area is provided by O'Sullivan in her chapter describing the development and impact of a 1940s/1950s nostalgia room at Newham University Hospital. The deep emotional responses evoked in patients, relatives and clinicians using this multisensory, interactive space are clear from the stories recounted in this thought-provoking chapter. Part IV ends with a chapter by Selai recommending a methodological approach for measuring health status, well-being and quality of life, providing a valuable toolkit for future research. Notwithstanding the need for continued research in this area, there is enormous potential for museums to play a significant and important role in social and health care in the future.

In the final part of this book, authors recount experiences of handling sessions involving under-represented audiences such as diaspora, the socially excluded (such as prisoners) and visually impaired museum users (see chapters by Lynch and Samuels in this section). Weisen sets out a critique of European social and cultural policies that seek to encourage museums to provide object handling for visually and mobility impaired visitors. It presents some of the significant barriers that many disabled people face in the enjoyment of museum collections and an assessment of museums' successes to date in adopting these recommendations. Irrespective of the progress achieved in some museums, this chapter highlights the size of the challenge remaining in order for the cultural rights of disabled and visually impaired people to be recognized.

The case studies, stories, research and examples presented in this book convey a powerful message regarding the value, possibilities and importance of tactile provision in museums. The experiences elicited by touch in this context go beyond, but do not exclude, learning and enjoyment to include deep emotional responses stimulated by object handling. Although this volume does not directly address issues such as conservation and preservation, which are often high on the priority list of museums, the chapters remind us of the importance of remembering the audience and considering their needs. This is particularly highlighted for under-represented audiences, such as those with visual impairments. The research discussed in this book provides useful, practical, guidelines for museums – for example, the importance of providing tactile pictorial information, since this improves visually impaired visitors' memory for textual information.

In bringing together scientists, museologists, scholars, clinicians and curators, this volume provides an exciting opportunity to think creatively about touch and object handling and the provision of tactile experiences that cater for the specific and varied needs of the diverse museum visitor. Another key outcome of this volume is the identification of a series of strands of further research, particularly regarding the therapeutic potential of object handling and the role of touch in knowledge transfer. Finally, the book demonstrates the enormous potential museums can play in learning, enjoyment, health and social care, centred around a multisensory approach to tactile provision.

Acknowledgements

I am extremely grateful to Sally MacDonald (Director of Museums and Collections, University College London), David Prytherch (Artist and Research Fellow in Haptics and Computer Interfaces, Birmingham Institute of Art and Design) and Guy Noble (Arts Curator, University College London Hospitals) who, in their capacity as consultants, provided invaluable editorial input on the development of this volume.

Refereences

Candlin, F. (2003), 'Blindness, art and exclusion in museums and galleries', *Journal of Art and Design Education* 22: 100–10.

Candlin, F. (2004), 'Don't touch! Hands off! Art, blindness and the conservation of expertise', *Body and Society* 10: 71–90.

Candlin, F. (2006), 'The dubious inheritance of touch: Art history and museum access', *Journal of Visual Culture* 5: 137–54.

Classen, C. (2005), 'Touch in the museum', in C. Classen (ed.), *The Book of Touch*, Oxford: Berg, pp. 275–86.

Classen, C. and Howes, D. (2006), 'The museum as sensescape: Western sensibilities and indigenous artefacts', in E. Edwards, C. Gosden and R. Phillips (eds), *Sensible Objects: Colonialism, Museums and Material Culture*, Oxford: Berg, pp. 192–22.

Heller, M.A. and Schiff, W. (1991), *The Psychology of Touch*, Hillsdale, NJ: Lawrence.

Pye, E. (2008), *The Power of Touch: Handling Objects in Museums and Heritage Contexts*, Walnut Creek, CA: Left Coast Press.

Part I
What Do We Mean by Touch?

–1–

Museums, Modernity and the Class Politics of Touching Objects

Fiona Candlin

There is some considerable evidence that visitors to the first public museums touched the collections. David Howes and Constance Classen have suggested that this had stopped by the early to mid-nineteenth century, a change that they attribute to modern notions of observation, objective science, surveillance techniques, the visual display of capitalism and the increasing equation of touch with 'primitive cultures'. This is an excellent analysis but what it forgets is the degree to which class inflects touch. A closer analysis reveals that the upper classes always had licence to touch and their touch was deemed rational and non-damaging. On the few occasions when the lower classes could touch museum collections it was considered unruly and dirty. Moreover, in many cases their admittance to museums was actively restricted. It was only when museums became properly public and open to all classes rather than just to an elite, that touch was prohibited. The question remains: is a similar dynamic still in place today? Is an elite touch still considered rational? Is the touch of the masses still filthy?

Introduction

Throughout the twentieth century numerous writers paid close attention to concepts of vision and the forms of knowledge that vision enables. In doing so they have tended to specify vision as the dominant sensory mode of modernity and to relegate the other senses, particularly touch, to a less sophisticated past. For instance, within art history Erwin Panofsky constructed a trajectory from classical art, which 'recognized only what was tangible', to the visual and conceptual refinement of Renaissance painting (Panofsky 1991, p. 41). His near-contemporary Clement Greenberg argued that the arts had conflated different mediums and their correlative senses until modernist artists finally perfected painting and sculpture which had purely visual effects (Greenberg 1985, p. 43). Despite their marked differences both authors considered this visually oriented, modern approach to be both aesthetically and intellectually superior to the earlier multisensory forms of practice. A similar pattern can also be discerned in other disciplines; the anthropologist Walter Ong suggested that inventions such as the printing press, microscope and telescope were keys to the transition from oral to visual cultures (Ong 1967), as did Marshall McLuhan (McLuhan and Fiore 1967). Even writers such as Martin Jay, who are highly critical of homogenous and hegemonic notions of vision, have still accepted that sight is 'the master sense of the modern era' and have paid little attention to the ways in which the other senses may also have had a role in the construction of modern knowledge, art or subjectivity (Jay 1988, p. 3).

Museum studies have proved no exception to this ocularcentric tendency, indeed they might be one of the main exponents of it. Perhaps most significantly for the discipline, Michel Foucault demonstrated the shift from pre-modern and multisensory to modern and visual knowledge in relation to cataloguing and collecting. He described how specimens from archives, gardens, herbariums and natural history collections had come increasingly to be laid out alongside each other forming tables that could literally be read off by a viewer; in short, collections were one of the places where modern visual systems of knowledge were generated (Foucault 1970). More recently Donald Preziosi has developed Foucault's thesis arguing that the modern museum 'was the most extraordinary optical instrument of all; the veritable summa of opticality, of visuality' (Preziosi 1996, p. 106). As such, the museum was not only productive of knowledge about objects but functioned as 'an instrument for the manufacture of ... societies, ethnicities, races, classes, genders, and individuals; of history, progress, and moralities' (ibid.). For Preziosi it is the visual structure and dynamics of the museum that has made it such a powerful contributor to the formation of modern subjectivity at both an individual and collective level.

This exclusive association between vision, knowledge and modern subjectivity potentially presents problems for the collective project represented in this book – namely, to conceptualize the use of object handling within museum education provision and beyond. After all, how can touch tours or object handling offer any kind of educative experience if touch is not considered to make any contribution to modern learning or subjectivity? Rather than suggesting that museum educators abandon all forms of tactile provision, this chapter reconsiders the accepted model of modern, visually oriented learning. In particular, I question whether there was a wholesale transition from tactile, pre-modern sensibility to visual modern knowledge, and whether public museums, as paradigmatically modern institutions, have always structured learning in relation to visual experience.

Touching and Rationality in the Eighteenth-century Museum

One of the few accounts to provide an alternative to ocularcentric notions of the museum is 'The Museum as Sensescape: Western Sensibilities and Indigenous Artefacts' by Constance Classen and David Howes. In this fascinating article, Classen and Howes draw on eighteenth-century journals and diaries to describe how museum visitors of the period regularly handled the objects on display. They note that in 1702, on a visit to the Ashmolean Museum, Celia Fiennes contrasted the light weight of a cane to its solid appearance and picked up loadstones, moving them around to explore the effects of magnetism. Similarly, they quote Sophie de la Roche who, at the other end of the eighteenth century, handled the collections at the British Museum:

> With what sensations one handles a Carthaginian helmet excavated near Capua, household utensils from Hercaulaneum ... There are mirrors too, belonging to Roman matrons ... with one of these mirrors in my hand I looked amongst the urns ... Nor could I restrain my desire to touch the ashes of an urn on which a female figure was being mourned. I felt it gently, with great feeling. (Classen and Howes 2006, p. 202)

The authors argue that at this point in time touch was considered to provide an entirely legitimate way of learning about and enjoying the museum collections but by the mid-nineteenth century, the acceptance of tactile experience had disappeared. Here they cite the renowned art writer Anna Jameson, who in 1841 expressed her disapprobation of touching collections in the strongest terms:

> We can all remember the public days at the Grosvenor Gallery and Bridgewater House, we can remember the loiterers and loungers ... people who instead of moving among the wonders and beauties ... with reverence and gratitude, strutted about as if they had a right to be there; talking; flirting; touching the ornaments – and even the pictures. (ibid.)

Classen and Howes argue that this sensory shift was due to the development of industrial capitalism which emphasized the appearance of commodities, surveillance (particularly within the context of social institutions), and to the growing use of visualizing techniques in science.

In maintaining that touch did contribute to museum visitors' knowledge, imaginative experience and subjectivity Classen and Howes implicitly challenge Foucault and Preziosi's view that touch had no intellectual role in museums during the eighteenth century. At the same time they still concur with the dominant supposition that touch was eventually excluded from museums. Effectively, then, Classen and Howes also posit a transition from multisensory to visual learning within museums; they move the date of modernity forwards rather than specifically challenging the accepted link between vision and modernity. That transition and the link between vision and modernity can, however, be questioned if we reconsider Classen and Howes's sources from the standpoint of *who* touched rather than *when* they touched.

Celia Fiennes was an aristocrat who investigated the Ashmolean collection in 1702 on one of her long journeys that eventually stretched as far afield as Newcastle and Penzance. Discussing these travels, Fiennes's biographer Christopher Morris comments on the privileges afforded to the aristocracy of this period recounting that, along the way, Fiennes went hare-coursing without permission, an activity which prompts Morris to comment that 'only someone of the most assured social position could have risked this defiance of the new and savage game laws' (Morris 1995, p. 23). Morris notes that aristocratic immunity from censure or punishment was so extensive at this time that 'when Celia's kinsman the fat Earl of Lincoln made his servants beat a "prentice boy to death for gazing at him in the street" the coroner's jury found that the boy "long before was sick of consumption and died of that disease" '(ibid). In this context it is highly unlikely that the museum assistants would have objected to Celia handling the loadstone and canes.

Similarly, not only was Sophie de la Roche wealthy enough to take the Grand Tour but she was extremely well connected; her stay in London included meeting the royal family and socializing with Warren Hastings, the Governor General of India and his family. De la Roche's social standing was important because it meant that she could easily gain access to the British Museum collection. This was much more difficult than we might imagine today.

In the late eighteenth century the British Museum was only open to the public on Mondays, Wednesdays and Fridays, with Tuesdays and Thursdays reserved for artists and private visits. It was closed at the weekends and on public holidays, thereby prohibiting visits by anyone in regular employment. In addition, acquiring a ticket was a labyrinthine and exclusive process: potential visitors had to apply in writing, giving details of their status and residence, to be approved and granted entry by the Principal Librarian who was the most senior member of staff. The applicant then had to return to collect their ticket, which could rarely be used on the day of collection. Moreover, as only ten tickets were allowed for each hour of opening, an immense backlog of applicants soon formed and the museum was forced to advertise announcing 'Persons applying are requested to send weekly to the porter to know how near they are upon the List' (Wilson 2002, p. 38). Entrance to the museum therefore required a minimum of three visits: application, collection and attendance, and preferably the use of a servant to regularly check the list. Anyone without independent means would scarcely have found the time to even acquire a ticket.

As an extremely well-connected visitor of means, Sophie de la Roche visited the British Museum on a private day and was given a personal tour by one of the curators. She therefore had plenty of time to view and handle the objects at her leisure. Her experiences form a sharp contrast with those recorded by the Birmingham merchant William Hutton who visited the year previously in 1785. Hutton had eagerly anticipated his visit to the museum but in order to gain access he had to pay a ticket tout the exorbitant charge of two shillings and, this being a public day, he was rushed around at great pace. Instead of being able to 'regale the senses, for two hours, upon striking objects', he compared himself to Tantalus who in Greek myth was condemned to reach for food and water that was perpetually snatched from his grasp (Hutton 1785: 191–2).

Fiennes, de la Roche and other elite visitors of the period all took their licence to handle museum collections for granted. As Classen and Howes note touch *was* understood as a legitimate and even essential means of engaging with art and artefacts during the eighteenth century. However, not everyone had the chance to do so; restricted opening hours, public and private days as well as the private nature of most collections meant that the working classes and, to some extent, even the middle classes did not gain access anywhere near as easily as the elite.

Equally importantly, touching objects in collections was not considered to be a universally valid activity. On the occasions that the working classes did have the opportunity to handle the museum collections, it was understood rather differently to when the elite did so. Sophie de la Roche happily picked up the weighty antique shields at the Tower of London but utterly disapproved of the female warder 'handling the things, turning them round and putting them back again', writing that 'it seems impossible that a woman, furthermore so ungainly in appearance, should be put in charge of pure gold and all that a crown implies' (de la Roche 1933, p. 129). Similar attitudes are also in evidence earlier in the eighteenth century. Like Celia Fiennes, the German traveller and eminent diarist Zacharias Conrad von Uffenbach also visited the Ashmolean Museum and the Royal Society Collection where he handled the objects, comparing weights, textures and the effects of magnetism. Predictably this did not stop him making ironic comments about the liberalism of the doormen and the surprisingly good condition of the collections, given that 'even the women are allowed up here for sixpence; they run here and there grabbing everything' (Uffenbach 1928, p.31). For Uffenbach, Fiennes and de la Roche, object handling is a way of investigating the empirical properties of the collections or of making imaginative connections with the past and in all cases their experiences are reported soberly and in some detail. Whereas they considered their propensity to touch as legitimate, they thought that the touch of their social inferiors was unruly and destructive. Indeed, they even considered that the close proximity of working people and valuable objects was enough to pollute or damage the objects. Both the possibility of touching objects and the recognition that touch could contribute to learning, pleasure and subjectivity all depended on *who* touched.

Thus, rather than arguing that knowledge in the eighteenth century was structured through vision or remained multisensory it makes better sense to think about for whom is it primarily visual, tactile or both, or in what ways is sensory experience variegated by social standing.

Curtailing Touching in the Nineteenth-century Museum

Still, the fact remains that museums gradually became spaces in which visual experience was apparently prioritized over tactility. Arguably, then, the elite eighteenth-century visitors' tactile encounters could simply be a hangover from an earlier period, a remnant of the patterns of learning connected to aristocratic personal ownership. As Classen and Howes suggest, it may have taken until the nineteenth century for museums to become modern visual spaces. Accordingly, this section considers how class and variegated conceptions of touch were played out in the nineteenth-century museum.

One of the most significant institutional changes between eighteenth- and early-mid-nineteenth century English museums concerned the type of visitors. In 1799 Joseph Planta became principal librarian at the British Museum and he both extended the opening hours and made tickets available for same-day admittance. Nine years later, in 1808, Planta finally managed to persuade the trustees to abolish the ticketing system altogether. Visitors still had to sign in and were shown round by guides, but after a further two years this system was also changed and for the first time people could wander for as long as they liked through the museum (Wilson 2002, p. 99). Not surprisingly, the rise in visitor numbers was noticeable, increasing from 13,406 visitors in 1807–8 to 40,500 by 1817 (Miller 1974: 96, 107n. 3).

Despite these increases several Government Select Committees on Museums of the 1830s and 1840s asked searching questions about access to museums and public monuments. Although some curators put up considerable resistance to widening access, there was considerable pressure from reformers who saw museums as having a much broader social role, particularly in educating the working classes. The Select Committee on Arts and Manufactures of 1835 repeatedly heard its invited witnesses attribute the superiority and market-dominance of French craftsmanship to the numerous public galleries in France. Several witnesses and some of the committee members thought that English museums should also work to develop the workers' aesthetic sensibilities through museums and thereby improve the standards of manufacture and boost the economy (Select Committee on Arts and Manufactures 1835, pp. 201, 219, 576).[1] Additionally, it was felt, museums would offer an alternative to the gin palace (Select Committee on Condition &c. of the British Museum 1836, p. 3408). The reformers won the day and in 1836, for the first time, public museums started to open over the holiday periods. The effect was dramatic. During the Easter week alone of 1837, 36,223 visitors went to the British Museum, approximately three times the number admitted in the whole year that Sophie de la Roche visited. The year after, in 1838, the National Gallery moved from its cramped Pall Mall apartments into new purpose-built premises in Trafalgar Square and 397,649 people visited, rising to over 500,000 by 1840.

This dramatic change in levels of access for working-class visitors had important consequences for sensory modes of learning and pleasure on offer within the museum. Not least, the abolition of private days meant that there was little segregation between elite visitors and the general public. It was therefore impossible to maintain two different levels of physical access to the collections as had been the case in the eighteenth century when elite visitors handled objects at leisure and ordinary visitors were rushed through. Clearly, with hundreds of thousands of people visiting on more-or-less the same terms it was impossible to allow everyone to touch the exhibits but, at the same time, it was clear that contemporary anxieties about conservation concerned the physical presence of the working classes rather than being purely a response to the expansion in visitor numbers.

For example, Anne Jameson's disapprobation of visitors touching the furniture and 'even the paintings', which Classen and Howes cite as marking a distinctive change in attitudes to touch, is explicitly directed at people below her own class. In the passage quoted earlier Jameson is writing against art being made accessible to the masses:

> I have heard the wish expressed, that these treasures were assembled in one place – in one national gallery, *easily and constantly accessible to all.* I cannot say I sympathize with the wish. (Jameson 1844: xxxiv)

For Jameson, keeping art in private houses effectively barred 'the loiterers and loungers, the vulgar starers' and those vulgar visitors who talk, flirt and touch the ornaments, 'as if they had the right' (ibid). It is not the activity of touching objects that is disgusting but the kind of people who will be potentially permitted to touch the objects.

These sentiments were echoed repeatedly elsewhere, perhaps most explicitly in the 1850 Select Committee Report on the National Gallery where the cleanliness of paintings was a major concern. The report notes that in the vicinity of the National Gallery were several large chimneys belonging to baths and washhouses: that the Trafalgar Square fountains were powered by steam-engines emitting volumes of smoke, as did the chimneys of the nearby club-houses, and that there was constant steam-boat traffic on the nearby stretch of the Thames. Even so, several witnesses to the Committee opined that the working-class visitors were the cause of the damage done to the paintings: 'there is a necessity to protect (the paintings) from the very bad atmosphere, and the quantity of dust and dirt that rises up, *open as it is to all classes*' (Select Committee on the National Gallery 1850, p. 619, my emphasis) and there was serious discussion about making the museum less physically accessible to the general public. Even the more tempered final report from the Select Committee still attributed damage to the visitors rather than to environmental pollution (Select Committee on the National Gallery 1850: ii).

During the nineteenth century the working classes were given increased access to museums, but it was not on the same terms as their elite predecessors. Once museum visitors came from across 'all classes', a corresponding shift occurred in the level of contact they were permitted to have with the collections. One of the costs of improved public access to museums was the loss of touch as a valid means of engaging with the collections:

> [W]hether [the visitor] moves in the ranks of the rich and influential or wears a humble fustian jacket, he finds this free ingress equally afforded: all he has to do is to show

> the common sense and the common justice of respecting and leaving unharmed and untouched the things which he sees. (Land 1847–50: 37)

Class and the partial democratization of the museum was one of the major factors in the shift towards a visually oriented museum and in changing attitudes towards object handling. Hypothetically, the Victorians could have recognized the value of object handling while simultaneously acknowledging that for practical reasons it had to be curtailed. Instead, once the working classes started visiting museums, touching objects was no longer regarded as an educative or pleasurable experience; it now ran against 'common sense' and 'common justice'. Crucially, however, as the next section shows, this was not the case for everyone.

Object Handling Behind the Scenes

During the nineteenth century the combined expansion of museum visiting and the negative estimation of the working class meant that object handling was forbidden in the galleries. Yet it is important not to view this as a final transition from a multisensory to a visually oriented museum for, while museum visitors were forbidden to touch the collections from the early nineteenth century onwards, object handling remained highly valued among experts.

During the late nineteenth and early twentieth centuries there was a great deal of discussion about the role and training of curators and it is clear from early Museum Association publications that curators were expected to have a close tactile involvement with their collections. Speaking on 'The Education of the Curator' in 1894, J. Paton, curator of Kelvingrove museum in Glasgow, recommended a system of apprenticeships to instruct young men 'in all the details of museum work', namely, 'the identification, labelling and registration of specimens, in methods of preparation and mounting, in the mysteries of preserving and cleaning, and in short in all the innumerable details of museum work' (Paton 1895; 102). Early twentieth-century training courses instructed curators in an even wider range of manual skills – making cabinets, panoramas and other mechanisms for display, gutting, skinning and preserving animals, rock abrasion and paper conservation. As a contributor to the *Museums Journal* of 1930 pointed out, curating involved both mental *and* bodily skills (ibid.: 293).

Importantly, these practical skills were considered to be inextricable from the intellectual and creative aspects of curating. In the late nineteenth and early to mid-twentieth century most curators were connoisseurs and, as Eugene Kleinbauer notes in *Modern Perspectives in Art History*, 'the tasks confronting the connoisseur are demanding and exacting. Above all, they require long, intimate experience with original works of art. Sound connoisseurs are not armchair wizards who go about their business by examining photographs, but scholars who handle and observe critically the work of art' (Kleinbauer 1989, p. 43). The importance of this physical engagement with objects is made very clear in curators' autobiographies and letters. John Pope-Hennessy mentions that, as a curator at the Victoria and Albert museum in the early 1940s, he personally positioned the objects in their cases and that on one occasion when he was dissatisfied with his arrangement of medieval art works he asked Leigh Ashton, the director, to undertake it instead (Pope-Hennessey 1991, pp. 162–3). He then goes on to note that 'you cannot know about works of art unless you live with them and this, in a modest fashion, I have done' (Pope-Hennessey 1991, p. 89). Similarly, his contemporary James Laver recounts his own early immersion in the prints and drawings department where 'by dint of turning over our own prints I gradually came to distinguish good from bad' (Laver 1963, p. 87). It was this intensive contact with works of art which then enabled him to identify new art works and prints, and to say 'with the finality of conviction, "No; it is a late impression from a very worn plate"' (ibid.).

In the context of connoisseurship, object handling was closely allied to the authentification of art works and artefacts, but contemporaneous art historians also articulated a connection between touching and aesthetic response. In the 1930s, Adrian Stokes developed John Ruskin's notions of the artist's expressive touch to suggest that that touch constituted a means of appreciating and understanding art work. Here Stokes describes Henry Moore's *Composition 7* which formed part of his exhibition at the Leicester Galleries in 1933:

> Mr Moore has cut and polished this stone to display a silvery light and to offer the swift, greased smoothness of metals; yet it is most lovely to touch. The slight protuberances, the gradual, unsharpened ridges upon the major forms, the swift convergences that the eye apprehends immediately but around which the hand hesitates, exploring further and further, are as a solid cream upon the surface of this stone, a cream whose clefts are wrinkles – wide, shallow and smooth. (Stokes 1978: 311)

Two decades later, Herbert Read, who gave Pope-Hennessy his first job writing reviews for *Burlington Magazine*, argued that sculpture is a tangible, palpable art which pertains to volume, texture, mass and materiality; all of which can best be understood by touching it. For Read, it is only when sculpture engages with tactility rather than vision that it 'attains its highest and its unique aesthetic values' (Read 1956: 70).

However, the possibility of learning or enjoying sculpture's tangible qualities remained highly demarcated by status; as Clement Greenberg pointed in response to Read's *Art of Sculpture*, almost nobody has the opportunity to touch sculptures and, we might add, certainly not with the impunity that Read and Stokes possessed (Greenberg 1993: 272). Their status as eminent art critics, and to a lesser extent as artists, gave them privileged access to objects, something that was also obviously shared by the curators. Pope-Hennessy may have referred to his contact with objects as 'modest', but as director of first the Victoria and Albert Museum and later the British Museum there can be few people who have spent their lives among such astonishingly rich collections.

Moreover, the exclusivity of rational and aesthetic object handling was further underlined by the wider precepts of connoisseurship. For connoisseurs the identification, authentification and aesthetic appreciation of an art work or artefact was predicated upon a consideration of its formal and material qualities. This object-based approach edited out any consideration of the social factors involved in its production, interpretation or valuation and thereby implied that Western aesthetic tastes were universal. Effectively, then, connoisseurship denigrated any other standards of taste or judgement while simultaneously supporting a lucrative art market from which many connoisseurs personally benefited.

Possibilities for Object Handling in Contemporary Museums

Clearly, then, there has not been a wholesale transition from pre-modern multi-sensory forms of learning to modern visually oriented learning. Rather, the valuation of tactile experience has always depended upon who was touching the collections and in what capacity. Where, then, does this leave the contemporary education curator who is introducing tactile provision as part of access programming in a contemporary museum?

In some ways the history of object handling is extremely problematic for its use in a modern educational context. While some museums have always kept handling collections, mainly for schools, these often contained unprovenanced objects or items of lesser quality than the museum's main collections. Opportunities to handle objects from within museum collections have almost invariably been predicated upon an extremely privileged level of access to museum collections, and simply changing who is allowed to touch will not remedy this legacy since, historically, working-class visitors have not been credited with the ability to learn rationally through touch. Thus the success of tactile access provision depends upon there being a clear understanding of how touching enables *all* museum visitors and non-experts to learn about the objects at hand. This, in turn, presents further problems because connoisseurship provides our only sustained model of object handling in relation to art work and museum collections. Given the deeply exclusive nature of connoisseurship it may not be the best precedent to adopt for the modern public museum.

At the same time, however, the variegated history of object handling, including connoisseurship, demonstrates that touching the objects in museums is not inherently pre-modern or simplistic; on the contrary, it has contributed to both the visitors' subjective experience, (for instance, Sophie de la Roche's sense of her relation to classical history) and to expert knowledge. These precedents would then suggest that object handling can also be a personally productive and educative experience for today's museum visitors.

Thus, object handling as part of contemporary access programmes might mean that the established dichotomy between elitist, connoisseurial approaches and contemporary contextual and inclusive approaches to art objects has to be rethought. Historically, connoisseurship involved a tactile engagement with materiality, weight, texture, form and manufacture, and an appreciation of how all these aspects coalesce into an object of greater or lesser complexity. As such, connoisseurship potentially provides a starting point for education curators to begin exploring the uses and possibilities of object handling. These techniques may previously have been linked to exclusive Western practices of aesthetic pleasure, but there is no reason why that has to remain the case. Instead, the elite handling skills of Fry, Ashton and Pope-Hennessey could be disinterred and could start to inform the critical, contextualized interpretation of objects. Connoisseurship and access programming may be an unlikely alliance but it may ultimately be a productive one.

Note

1. All citations from Select Committee reports refer to the numbered lines of evidence and not to page numbers.

References

Classen, C. and Howes, D. (2006), 'The museum as sensescape: Western sensibilities and indigenous artefacts', in E. Edwards, C. Gosden and R. Phillips (eds), *Sensible Objects: Colonialism, Museums and Material Culture*, Oxford: Berg, pp. 192–222.

de la Roche, Sophie (1933), *Sophie in London 1786: Being the Diary of Sophie de la Roche*, London: Jonathan Cape.

Foucault, M. (1970), *The Order of Things. An Archaeology of the Human Sciences*, London: Tavistock Publications.

Greenberg, C. (1985), 'Towards a Newer Laocoon', in F. Frascina (ed.) *Pollock and After, The Critical Debate*, London: Harper & Row.

Greenberg, C. (1993), 'Roundness isn't all: Review of *The Art of Sculpture* by Herbert Read', in *Clement Greenberg: The Collected Essays and Criticism*, Chicago/London: John O'Brian.

Hutton, W. (1785) *Journey from Birmingham to London*, Birmingham: Pearson & Rollason.

Jameson, A. (1844), *Companion to the Most Celebrated Private Galleries of Art in London*, London: Saunders & Otley.

Jay, M. (1988) 'Scopic Regimes of Modernity', in H. Foster (ed.) *Vision and Visuality*, Bay Press.

Kleinbauer, E. (1989), *Modern Perspectives in Western Art History: An Anthology of Twentieth-Century Writings on the Visual Arts*, Toronto/Buffalo: University of Toronto Press.

Land, J.T. (1847–50), 'The British Museum', in *The Land We Live In: Pictorial and Literary Sketch Book of the British Empire*, Vol. I, London: Charles Knight.

Laver, J. (1963), *Museum Piece or the Education of an Iconographer,* London: Andre Deutsch.

McLuhan, M. and Fiore, Q. (1967), *The Medium is the Message*, Harmondsworth: Penguin Books.

Miller, E. (1974), *That Noble Cabinet: A History of the British Museum*, Athens OH: Ohio University Press.

Morris, C. (ed.) (1995), *The Illustrated Journeys of Celia Fiennes 1685–c.1712*, Stroud: Sutton Publishing.

Ong, W.J. (1967), *The Presence of the Word: Some Prolegomena for Cultural and Religious History*, New Haven, CT: Yale University Press.

Panofsky, E. (1991), *Perspective as Symbolic Form*, New York: Zone Books.

Paton, J. (1895), 'The education of the curator', in E. Howarth and H.M. Platnauer (eds) *Report of Proceedings with Papers Read at the Fifth Annual General Meeting of the Sheffield and York Museum Association.*

Pearce, S. (2003), 'Introduction', *Lives of the Founders of the British Museum*, London: Routledge.

Pope-Hennessey, J. (1991), *Learning to Look*, New York: Doubleday.

Preziosi, D. (1996), 'Brain of the earth's body: Museums and the framing of modernity', in P. Duro (ed.), *The Rhetoric of the Frame: Essays on the Boundaries of the Art work*, Cambridge/New York: Cambridge University Press.

Read, H. (1956), *The Art of Sculpture*, Princeton: Princeton University Press.

Select Committee on Arts and Manufactures (1836), Report with Minutes of Evidence, London.

Select Committee on Condition &c. of the British Museum (1836), Report with Minutes of Evidence and Appendix, London.

Select Committee on the National Gallery (1850), Report with the Minutes of Evidence and Appendix, London.

Stokes, A. (1978), 'Mr Henry Moore's sculpture', in L. Gowing (ed.) *The Critical Writings of Adrian Stokes 1930–1937*, London: Thames & Hudson.

Uffenbach, Z.C.V. (1928), *Oxford in 1710*, Oxford: Blackwell.

Wilson, D.M. (2002), *The British Museum: A History*, London: British Museum Press.

–2–

Making Sense of Touch

Charles Spence and *Alberto Gallace*

'How can Anthony Caro's *Sculpture Two* or Canaletto's *Piazza San Marco* be made accessible to people who have never seen or who no longer see?' (Candlin 2003: 101). After more than a century of research by psychologists, psychophysicists and cognitive neuroscientists on the topic of tactile/haptic perception, we still do not have any concrete answers to this fundamental question. Nevertheless, the necessity of finding practical solutions to this problem has become all the more pressing for museum and gallery curators given the recent publication of the Disability Discrimination Act. Museums and art galleries in the UK must now, as a legal requirement, ensure that their facilities and exhibitions are made accessible to visitors with sensory (e.g. visual) impairment. In this chapter, we briefly discuss the key findings to have emerged from the scientific study of touch over the last few years. We also highlight some of the most important issues that will need to be addressed in the years to come if genuine progress is to be made in providing tangible solutions to the problem of how to make the exhibits in museums and galleries more accessible to the visually impaired. The likely insights derived from the research reviewed (and proposed) here should also help to enhance the experience of sighted individuals in museums as well.

Introduction

The last few years have seen a dramatic increase in the number of conferences and workshops that have been convened with the expressed aim of trying to understand how to make our museums and art galleries more accessible to those suffering from some sort of sensory impairment. These meetings have brought together individuals from many different backgrounds, including museum curators, educators, psychologists, neuroscientists, those with some form of sensory impairment (as well as those working on their behalf), not to mention artists and art historians. However, while everyone involved agrees that something urgently needs to be done in order to make museums and galleries more accessible (or 'touch-friendly', though see Candlin 2004), little progress has, as yet, actually been made in terms of providing concrete recommendations and/or insights with regard to how to improve the educational and/or aesthetic experience of those visiting museums and art galleries (Candlin 2003).

In this chapter, we critically assess recent scientific advances in the psychology, psychophysics and cognitive neuroscience of touch (and 'haptic' perception, i.e. active touch) with a view to evaluating their ability to provide concrete solutions to the question of how to enhance the multisensory experience of cultural objects – be they artifacts, pottery, tools, and/or sculpture – by those individuals with a sensory impairment (see Classen and Howes 2006). More specifically, we focus on the question of what can be done for people with impaired vision since this is the area where most of the research interest has been directed over the last few years. It should, however, be noted that the benefits of a more 'multisensory' approach to object handling and museum education are not restricted solely to those having some form of sensory impairment, but should also help curators and educators to enhance the experiences of the sighted museum visitor as well (e.g. Arnold 2003; MacGregor 1999; see also Minogue and Jones 2006, for a recent review of the role of haptics in education).

One of the central questions in this area of research is: what specific attributes of a museum object, artefact, or sculpture should one try to capture when creating a tactile replica (or model)? We believe that it is only by gaining a fuller understanding of the multisensory nature of our everyday perception and aesthetic appreciation of objects, and also of the dominance of certain tactile object attributes over others (i.e. shape, size, texture, etc.), that we will be able to make a more informed decision about what it is, exactly, that a tactile replica should be capturing or reproducing (see Classen 2005; Cranston 2003; Diaconu 2006). These then are some of the most important insights that the scientific study of touch/haptics will need to provide in the years to come if we are to create educationally – and/or aesthetically – meaningful tactile replicas, designed to be handled by museum or gallery visitors, no matter whether or not they happen to be visually impaired.

The Scientific Study of Touch

One realization made clear to us from our attendance at many of the meetings on touch and object handling over the last few years is the limited scope of much of the psychological and neuroscientific research published to date (including some of the research conducted in the Crossmodal Research Laboratory in Oxford). At present there appears to be a substantial gulf between the findings coming out of the many research laboratories around the world studying tactile perception in relation to the practical problems currently faced by museum and gallery curators: namely, how to make museum and gallery exhibits more accessible to the visually impaired visitor, especially given the obligations laid down by the Disability Discrimination Act.

It is, for example, surprising to see just how much of the psychophysical research on tactile sensation over the years has restricted itself solely to an assessment of people's ability to perceive and evaluate surface texture (i.e. the *microstructural* properties of an object or surface; see Lederman 1982, Lederman and Klatzky 2004, for reviews). Far less research has been devoted to the study of people's perception of the larger-scale *macrostructural* properties of objects, such as their size and shape. What is more, hardly anyone has attempted to investigate how the micro- and macro-geometric properties of an object or artefact interact in a laboratory setting.

An exception to this, however, is a recent study by Zampini et al. (2006) on the effects of handle texture on people's perception of the texture at the tips of deodorant sticks (Figure 2.1). The participants in this study had to hold a deodorant stick (which was either rough or smooth) in one hand while rubbing the cap of the deodorant (which had one of eight different roughness levels) on another part of their body (in this case, the contralateral forearm), and try to judge how rough the deodorant cap actually was. The results showed that people's perception of the roughness of the deodorant cap on the forearm was influenced by the roughness of the handle of the deodorant stick in the hand (for another exception, see Klatzky et al. 1989).

The majority of the laboratory-based research that has been published to date on the perception of surface texture has been restricted to assessing people's perception of sandpapers of varying roughness (or grit) levels. In hindsight, it seems somewhat bizarre that so much effort should have been devoted to studying people's perception

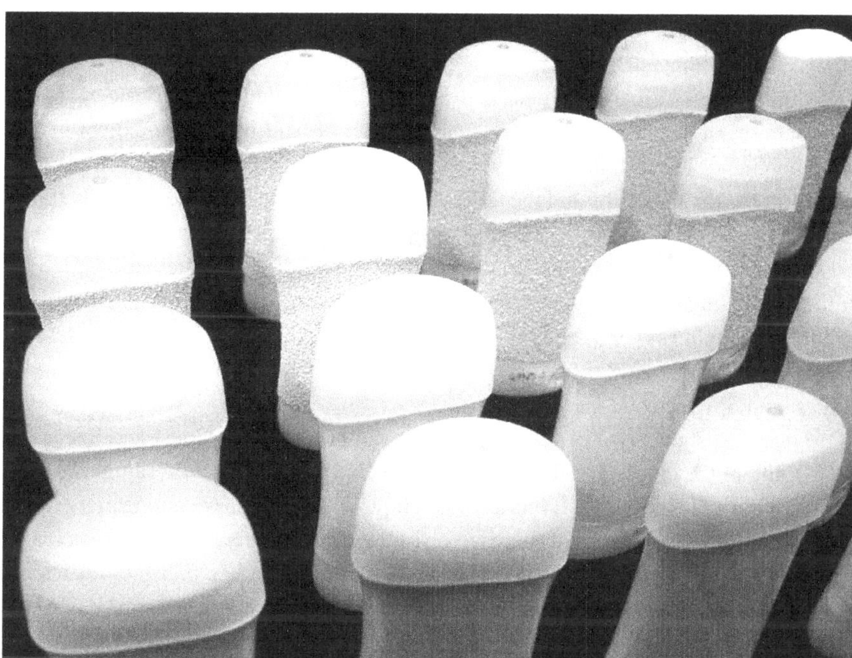

Figure 2.1 Deodorant sticks used in texture handling study (Zampini et al. 2006).

of a surface that they would rarely, if ever, be expected to touch in their everyday lives. But, it was not always so. In fact, the very earliest research into the perception of surface texture by Binns (1926) was actually designed to assess people's ability to discriminate between different finely graded cloths. Binns was particularly interested in understanding the relative reliability of vision and touch in terms of people's perception of the quality of woollen fabrics, a practical problem that is still relevant to us more than eighty years later. He was also fascinated by the question of whether or not the relative importance of the two senses would change as a function of an individual's level of expertise in assessing textiles (see below). Surprisingly, however, research on the perception of fabrics, and other materials/surface attributes, had more or less disappeared until recently (though see Yoshida 1968a, 1968b and 1968c), and it is only in the last five years or so that there has started to be something of a comeback (see Bergmann Tiest and Kappers 2006, 2007; Guest and Spence 2003a, 2003b, 2007).

For instance, Guest and Spence recently reported a number of studies in which they investigated the factors affecting people's perception of the roughness of pilled fabric samples. Their research demonstrated that while tactile cues played a more important role than visual cues in modulating people's perception of very fine (i.e. minimally pilled) surface textures, the reverse was true when the participants evaluated rougher (i.e. more pilled) fabrics (see Guest and Spence 2003a). Interestingly, a similar dominance of touch over vision was also reported by Heller (1989) when people had to evaluate the texture of a number of different Japanese abrasive sharpening stones (varying in grit value from 200 to 6,000).

In their latest research, Guest and Spence (2007) studied just how close the correspondence actually is between what people see and what they expect to feel with their hands. That is, they investigated whether there is a veridical match between the texture at the fingertips and the texture that a person expects to see visually. Interestingly, Guest and Spence found that while their participants were generally quite good at this cross-modal texture-matching task, they did show systematic variations from veridicality (see also Bergmann Tiest and Kappers 2007). That is, the degree of pilling (or roughness) of a seen fabric did not always provide an accurate impression of what that fabric would feel like when the participants subsequently touched it (Guest and Spence 2007).

Figure 2.2. shows two of the pilled fabric samples used by Guest and Spence in their study of the cross-modal visuo-tactile matching of pilled fabric samples (A is a 500-pill fabric; B is a 600-pill fabric). The graph (C) shows the relationship between the actual pill value of the fabric sample (in log units) that was felt by participants against the mean (visual) matched sample (log) pill estimates that they provided. Note that points above the main diagonal indicate smoother than veridical visual matches, whereas points below the main diagonal indicate rougher than veridical visual matches. Error bars show ±1SE.

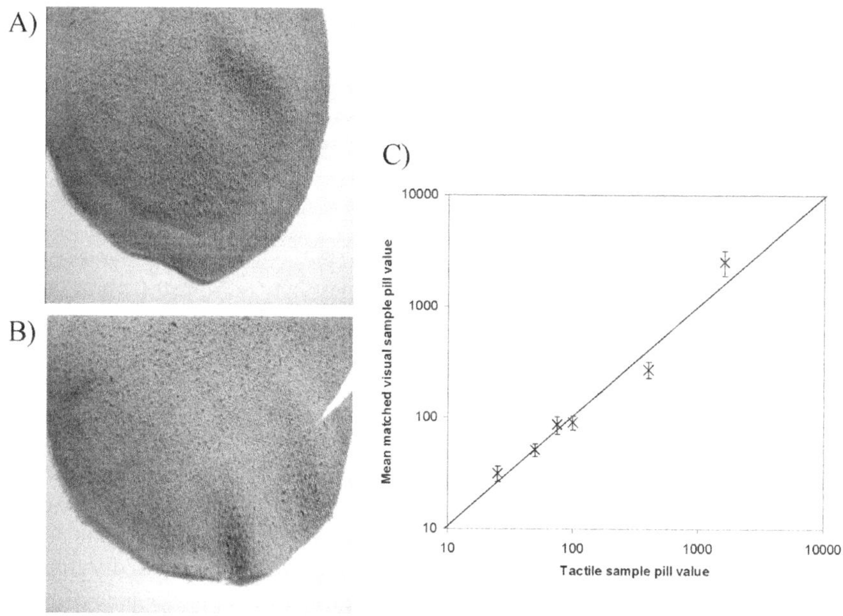

Figure 2.2 Pilled fabric samples used in cross-modal visuo-tactile matching study (Guest and Spence 2007).

It is interesting to note at this point that discrepancies between what an object feels like and what it looks like have also been reported in the context of archaeological studies of material culture – for example, see below MacGregor's (1999) sensory analysis of a selection of Scottish carved stone balls housed at the Kelvingrove Museum in Glasgow. The English traveller Celia Fiennes noted something similar when, on her visit to the Ashmolean Museum in Oxford in 1694, she picked up a cane 'which looks like a solid heavy thing but if you take it in your hands [is] as light as a feather' (Fiennes 1949: 33, cited in Classen 2005; see also Harvey 2002). A recent inspection of a number of contemporary design catalogues revealed the existence of a surprisingly large number of items that have been specifically designed to feel different from what one would have been led to expect simply by looking at them, and therefore generating a sense of surprise (see Schifferstein and Spence 2007). Similarly, the artist Miho Suganami (2003) also uses the incongruence between the senses of vision and touch in her artworks: for example, in the piece *Rosetta Stone III* (1997–1998) she has embedded a number of hidden light bulbs inside a hollowed piece of Hollington stone in order to make a material object (stone) that one would normally expect to feel cold, actually feel warm to the touch – see also her piece *Life* 2002 (for an overview of Suganami's work, see http://www.suganami.freeserve.co.uk/).

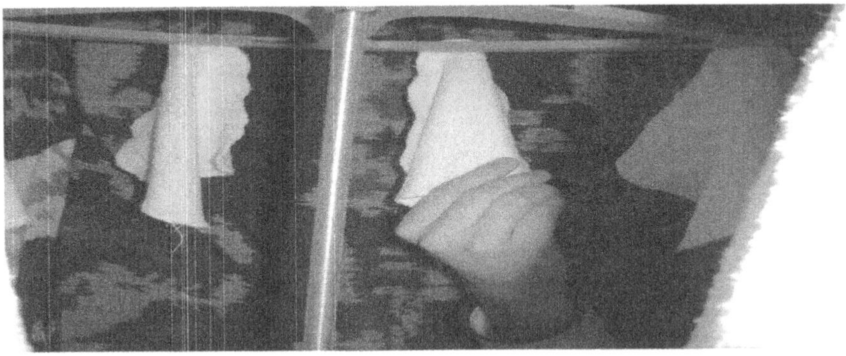

Figure 2.3 Softness evaluation study (Demattè et al. 2006).

Meanwhile, Demattè et al. (2006) have investigated some of the factors contributing to the perception of softness. The participants in their study had to evaluate the perceived softness of a series of fabric swatches using a naturalistic hand action (see Figure 2.3). The results showed that the participants were able to discriminate reliably between the softness of the various different fabric swatches which had been chemically treated in order to manipulate systematically their surface feel. More importantly, however, Demattè et al.'s results also showed that people rated the fabric swatches as being between 3 and 5 per cent softer when they were presented together with a 'pleasant' fragrance (lemon or lavender) than when presented with a somewhat less pleasant 'animal' odour (see also Laird 1932).

Interim Summary

Taken together, the results of the experiments reviewed so far – namely, Bergmann Tiest and Kappers 2007; Demattè et al. 2006; Guest and Spence 2003a, 2003b – clearly demonstrate the multisensory nature of our perception of the microgeometric attributes of haptically explored objects/surfaces. In other words, people's perception of the surface quality of an object or material reflects more than simply just what goes on at the level of their skin surface. In fact, visual, olfactory and even auditory cues – that is, the sounds that we 'hear' when we touch an object or surface (see Guest et al. 2002; Spence and Zampini 2006) – are all combined by our brains in order to give rise to our multisensory perception of the texture or 'feel' of a given surface (see also Candlin 2003).

It is worth noting at this point that people's perception of the macrogeometric attributes of objects are similarly influenced by the information available to the other senses (see Schifferstein and Spence 2007; Spence 2008; Warren and Rossano 1991, for reviews). Furthermore, any previous experience that people may have had with that type of object may affect their haptic exploration of, and/or their ability to recognize, the object concerned by touch alone (e.g. Klatzky et al. 1985; Lederman and Klatzky 1990).

Sensory Dominance

The latest psychophysical research has now started to quantify the relative contributions of each of our senses (primarily vision and touch thus far) to the overall multisensory impression that people have of the objects and surfaces with which they interact (see Ernst and Banks 2002; Ernst and Bülthoff 2004; Guest and Spence 2003b). The empirical evidence suggests that our brains tend to weight our multisensory perceptual experiences in favour of the most reliable of the individual sensory signals. Ernst and Banks used a haptic virtual reality set-up in order to demonstrate (using mathematical modelling) that a person's perception of the macrogeometric attributes of an object (such as the width of a rectangular bar) under normal viewing conditions depends more upon what that person sees than on what they feel. By contrast, a person's perceptual experiences will depend more upon what they feel (than on what they see) under conditions where the visual input has been degraded (see also Heller 1983, on the effects of blurred vision on sensory dominance). It should, however, be noted that while vision provides more accurate (or reliable) information about certain object properties, touch/haptics has been shown to provide more accurate information about others (e.g. Klatzky et al. 1993; Lederman et al. 1986; see Spence 2008, for a review). What is more, attention (i.e. what we choose to concentrate on) has also been shown to influence sensory dominance as well (Battaglia et al. 2003; Lederman and Abbott 1981).

Given such findings, one might have expected that those individuals with impaired vision would weight tactile information more highly than visual information (cf. Putzar et al. 2007). Similarly, one might also expect that those individuals who, through training (or for some other reason, cf. Craig 1977), have become more expert in the use of their hands – such as, for example, potters (Power and Graham 1976) or the expert assessors of woollen fabrics tested by Henry Binns[1] – would show an increased reliance on their sense of touch. However, the limited empirical research that has been conducted on this topic has so far failed to confirm this particular prediction (see Binns 1926; Power and Graham 1976).

Interestingly, the results of a recent neuroimaging study (using functional magnetic resonance imaging) has revealed some important differences in the patterns of brain activation between individuals who have different levels of tactile expertise (Saito et al. 2007). Saito et al. scanned the brains of two groups of individuals who were trying to discriminate haptically between different Mah-jong tiles (two-dimensional plastic tiles with various marks carved on one side) without being able to see them. One group of participants were new to the game of Mah-jong while the other group consisted of expert (i.e. highly-trained) Japanese players of the game who were able to identify the carved patterns by touch (with an accuracy of more than 80 per cent correct). Saito et al. observed no change in the activation of visual cortex when the participants who were new to the game felt the tiles (but could not see them). By contrast, extensive activation of visual cortex (in particular, of the primary visual cortex, V1, and the lateral occipital cortex, LO) was detected when the expert Mah-jong players were trying to decide whether two adjacent tiles were the same or not (known as a tactile matching task) using touch. It is interesting to note here that the activation of the visual cortex has also been reported when blind individuals read Braille or else explore objects haptically (for example, see Merabet et al. 2004, Sadato et al. 1996).

The recruitment of the visual cortex by the Mah-jong experts (but not by the naive participants) was also seen when the participants in Saito et al.'s (2007) study performed the haptic matching task on Braille characters, a type of tactile stimulus that none of the participants were familiar with. However, it is important to note that this additional activation in the 'experts' did not result in any enhanced behavioural performance of the tactile discrimination task. This latter null result is consistent with the null effects of haptic expertise reported by Power and Graham (1976), since they also assessed the sensory dominance of their potters using perspex blocks (i.e. with a stimulus that they were by no means expert in handling; see also Power 1981).

In the context of the present volume, it would be particularly interesting for future research to try and determine whether those who work more extensively with their hands, including sculptors (or those who work with sculpture) and those working in a material culture setting, would also show enhanced visual cortical activation when haptically interacting with the kinds of objects/stimuli on which their expertise is based (cf. Hall 1999; Kóbor et al. 2006; MacGregor 1999; Suganami 2003).

Interim Summary

While the scientific study of touch has traditionally been limited by its narrow focus on a small range of surface and object attributes, the research that is now being published is finally starting to look at a much broader (and more ecologically valid) range of tactile stimulus attributes, at both the micro- and macrogeometric scales. In terms of the museum and cultural heritage context, we believe that the recent scientific focus on the study of the multisensory nature of tactile perception helps to illustrate just how important the information available to *each* of our senses is in terms of its ability to affect our perception of the objects and artefacts with which we interact (see also Candlin 2003; Classen 2005; Classen and Howes 2006). These findings therefore help to emphasize the importance of catering for all of a museum or gallery visitor's senses, be it at the touch table, on a museum touch tour, or even in terms of the general multisensory features of the gallery space itself (see Losche 2006).

An Obsession with Sandpaper?

One of the principle reasons why scientists have spent so much of their time investigating people's perception of the roughness of sandpaper samples is that it is easy to *objectively* quantify the roughness of this kind of stimulus. That is, different sandpapers vary primarily in terms of their surface roughness (or grit size), and they thus make an ideal (i.e. well-controlled) laboratory stimulus for psychophysical investigation (see Bergmann Tiest and Kappers 2006). At the same time, this also makes it easier for scientists to replicate and extend the results of previous research. Furthermore, sandpaper can be easily purchased in a range of different roughness gradations. These factors have likely played an important role in accounting for the prevalence of research focusing on people's perception of (and ability to discriminate between) this particular kind of tactile stimulus.

The problem here, from a scientist's perspective, is that when one wants to study how people perceive more ecologically valid stimuli, it becomes difficult to discern exactly which of the many stimulus attributes that are physically changing are actually driving a participant's perceptual response. Thus scientists have (at least until recently) been severely hampered by their inability to characterize precisely how the naturally varying qualities of surfaces differ in terms of their physical characteristics. More importantly, they have struggled to show how those physical characteristics map onto (and influence) the psychological aspects of tactile/haptic perception.

Nevertheless, the latest research from Bergmann Tiest and Kappers (2006) has started to make tangible progress in terms of our understanding of the fundamental haptic/tactile dimensions underlying people's experience of various different realistic everyday materials. The authors conducted a study in which the blindfolded participants – none of whom had any special tactile/haptic expertise – had to rank 124 different material samples (from cardboard to carpet and steel mesh to different kinds of wood and tile – not forgetting the ubiquitous sandpaper) using a free sorting task in which the participants were asked to group the stimuli according to how similar they felt. The researchers then analysed the participants' responses using multidimensional scaling (MDS) techniques.

The results showed that there appeared to be just four key dimensions underlying people's haptic perception of materials. These dimensions included roughness and compressibility, as well as two others which were not so easy to provide meaningful verbal labels for. Interestingly, other studies that have used the MDS technique to look at the tactile perception of material qualities have typically come to the conclusion that between three and five key dimensions are needed to account for people's haptic perceptions. However, the particular dimensions themselves have varied between studies, and have included dimensions such as 'metallicness' versus 'fibrelikeness' (Yoshida 1968b), smooth/rough and hard/soft (Hollins et al. 1993, 2000), sticky/slippery (Hollins et al. 2000) and soft/harsh and thin/thick (Picard et al. 2003). One of the main problems then with this kind of research is that the number and nature of the dimensions of tactile sensation that a particular piece of material reveals appear to be somewhat dependent on the particular number and range of stimuli/materials that participants have to assess/group (see Bergmann Tiest and Kappers 2006, on this point).

Talking about Touch: The Need for a 'Lexicon of Touch'

In fact, the lack of a commonly agreed 'lexicon of touch' constitutes a very real hindrance to progress in this area. In other words, researchers do not have a recognized set of terms with which to describe the tactile sensations elicited by various material properties, although here have been sporadic attempts over the years to educate people's sense of touch, as well as their ability to describe or talk about their tactile experiences (see, for example, Hall 1999, Marinetti 1921, Montessori 1991). However, the paucity of terms that we currently have (or at least that we use regularly) to describe the nuances of our tactile experiences with objects is brought into sharp relief by contrasting them with the widely accepted lexicon of terms we use for describing visual and auditory stimuli such as colour and pitch (see Ember 1978).

An additional problem here relates to the fact that our tactile experience actually results from the operation of a number of different classes of neural receptors located in the skin that are in charge of transducing the various different aspects of tactile stimulation (for reviews, see Iggo 1977, McGlone et al. 2007, Rolls et al. 2003). These different classes of receptor are distributed idiosyncratically over the skin surface, and hence we have markedly different sensitivities to the various kinds of touch, as a function of the particular skin site that is stimulated (Bolanowski et al. 1999; McGlone et al. 2007; Weinstein 1968). To make matters worse, the various classes of receptors also have quite different temporal response profiles following the onset of tactile stimulation – for example, certain classes of receptors respond far more slowly than others. Furthermore, people do not even appear to be consciously aware of the stimulation of the receptors coding for pleasant touch that are located in the hairy, but not in the glabrous, skin (see McGlone et al. 2007 on this point).

The Cognitive Neuroscience of Tactile Perception

The spectacular developments in the field of cognitive neuroscience that have been seen over the last decade or so (see Cohen et al. 1996, Merabet et al. 2004, Sadato et al. 1996, Zangaladze et al. 1999, for just a few of the many examples) have also been visible at many, if not all, of the meetings and workshops on the topic of making museums more accessible. It has certainly been fascinating to watch the steady stream of neuroscience 'discoveries' illustrating just how much cortical plasticity there really is in the human brain and its impressive ability to adapt to its ever-changing environment – such as, for example, following early-onset blindness. The research in this area has now unequivocally demonstrates the important role that the visual, as well as the somatosensory, cortex plays in the tactile perception of both the blind (Sadato et al. 1996), and, more recently, also in the sighted (Merabet et al. 2004; Saito et al. 2007).

Cognitive neuroscientists have been able to develop a neural explanation for the enhanced tactile sensitivity that has sometimes been documented in the visually-impaired (e.g. Röder et al. 2004; though see Hollins 1989). In fact, such findings have even led some researchers to start questioning just how 'modality-specific' the primary sensory areas of the brain, such as the visual, somatosensory and auditory cortex, actually are (see Ghazanfar and Schroeder 2006). However, while such findings are certainly very important from a theoretical standpoint, it seems unclear (at least to us) what, if any, practical implications such findings provide in terms of our understanding of touch in a museum or cultural heritage context (cf. Heller 2004, on the limitations of neuroscience). As Candlin (2006: 151) puts it, we need to 'stop equating nicely coloured brain scans with an explanation of how touch functions or why it matters'.

Making a Tactile Replica

One way in which museum curators can make their exhibits more accessible to the blind and partially sighted is through the provision of tactile replicas that enable those who are unable to see to experience haptically something of the objects and artefacts that are on display. However, at present it is unclear exactly what aspects of an artefact or sculpture should be captured by a tactile replica. Is it the size that is important? Or is it the weight, the temperature (cf. Candlin 2004: 75; MacGregor 1999: 265; Suganami 2003), the texture, or the compliance of the material? We simply do not know. However, the results of Candlin's (2003) interview study of visually impaired museum visitors clearly highlights the inappropriateness of at least some of the current approaches. As Candlin notes, after discovering that the education curator responsible for access at one large gallery provided nothing more than a chunk of limestone for visitors to touch (given the prohibition on touching any of the gallery's collection of sculptures): 'Limestone in a three inch square does not convey the scale, weight, temperature, form, texture or rhythm of the work. These qualities are not irrelevant to the art object or to an understanding of what that art work is about' (Candlin 2003: 103).

This point about what is missed when people are only allowed to handle a piece of the material from which an object has been constructed is really brought home if one contrasts it with the richness of the tactile/haptic experiences that a person may experience if they are actually allowed to touch museum artefacts. Take, for example, MacGregor's (1999) haptic sensory analysis of a set of Scottish carved stone balls (CSBs) that he was allowed to touch:

> The first thing that you notice when your hands close around a CSB is their coolness to the touch. As the lateral motion of the fingers explores the form of the object, the texture rapidly becomes apparent … The initial enclosure of the hands rapidly established the sub-spherical shape of these objects and the exact form of them was rapidly determined by following their contours. In the case of the balls with a limited number of knobs, the number is readily and quickly established. When a larger number of knobs are present it becomes harder to establish the exact number by touch alone. (MacGregor 1999: 265)

There are clearly a number of problems with letting the general public touch artefacts, such as the deterioration of the objects that could be expected following heavy handling (Candlin 2004). Nevertheless, MacGregor's words still serve to highlight the richness of the temporal sequence of different tactile sensations that one may experience when actually exploring an artefact haptically, and this, presumably, is what we should be aiming for when trying to create a tactile replica.

The quotation (McGregor 1999) also nicely illustrates that, once one moves from the evaluation of sandpapers and chunks of limestone to the evaluation of real objects and artefacts, it suddenly becomes altogether unclear just exactly what the hierarchy of tactile sensations is. In other words, is there a dominant tactile/haptic sensory attribute when people pick up a particular object or artefact? Is it the object's weight, its size, surface texture, temperature, or some other attribute? Do different people experience a different hierarchy of tactile sensations, or are we all the same? And how do these sensations reveal themselves over time? How much of a role does the attentional focus of the person touching the object play here? As far as we are aware, scientists have not, as yet, been able to provide an answer to these intriguing questions.

In the future, then, it will be important for researchers to conduct further research in which they present a range of different objects to participants to evaluate which tactile sensations are dominant – that is, which sensations are perceived as being most salient in terms of perceptual experience – and, perhaps more importantly, what the hierarchy of dominance is.

Tactile Replicas: The Analogy with Visual Reproduction

It is perhaps enlightening here to consider the analogous situation involving the reproduction of visual images. Would Gericault's *The Raft of the Medusa*, for example, have the same impact on a viewer if seen as a reproduction of the painting in a book that might perhaps be only a twentieth of its actual size (i.e. 491 cm × 717 cm)? The answer is presumably 'no, it would not'. This observation leads to the suggestion that part of our aesthetic experience, at least of paintings, depends on their relative size (see Silvera et al. 2002 on this point). It will be important in the future to determine whether our perception of a tactile replica of a large piece of sculpture, for example, would be deleteriously affected to the same extent by the presentation of a small replica designed to fit snugly into the palm of the hand.

Another tricky problem associated with the creation of tactile replicas in the cultural heritage context is to determine how important it is for a tactile replica to reproduce the various materials involved in the construction of the original artefact. (Note also that different materials may be used in sculpture to communicate different feelings or emotions – see, for example, Hall 1999 and Marinetti 1921). One need only think here of how much of our appreciation of a painting (i.e. comprehending its meaning and emotional value), such as Vassilij Kandiskij's *Red-Yellow-Blue* or Karl Schmidt-Rottluff's *The Black House* is lost when it is reproduced in black and white (see Maffei and Fiorentini 1995). One might wonder whether the tactile/haptic equivalent here might be the comparison between an object made of many different materials and a replica made entirely from just one material.

Finally, it is important to note that the sound, smell (and perhaps, in former times, even the taste) of an artefact might be perceived as an integral part of a museum visitor's experience of an object – or at the very least influence their response to it (see Arnold 2003; Classen 2005, 2007; Hooke, 1705/1971, for historical perspectives). This problem becomes even more significant when one remembers that many of the objects displayed behind glass in museums and other cultural heritage contexts were originally made in order to be touched, held and/or actively used by people (see, for example, Classen and Howes 2006; Ouzman 2001). To what extent then should our tactile replicas be designed to mimic these non-tactile sensations associated with an object?

Sensory-discriminative versus Emotional Touch

Finally, in the years to come, researchers will need to focus far more of their efforts on questions related to defining which characteristics are important in determining whether people actually 'like' to touch a surface, or to hold and manipulate an object (see Major 1895 and McLoone 2003, for preliminary work on this topic in the context of the design of a 'haptically-pleasing' computer mouse). Returning to that chunk of limestone provided to visitors at Tate Liverpool (as described above), Candlin goes on to note:

> Moreover, the provision of small pieces of limestone or the equivalent completely ignores people's capacity for pleasure through touch. While the notion of aesthetic pleasure through touch is virtually unheard of within western metaphysics the interview respondents talked about touch in precisely these terms: 'You are sort of more aware of the beauty, when you can touch it'. (2003: 103)

Again this can be contrasted with the strong desire to touch/caress that many people have in response to particularly 'meaningful' objects, as illustrated by this quote from Edwards (1999: 227):

> This relic quality is perhaps most marked in photographic jewellery, where the physical trace – the photograph – is encapsulated with the bodily relic such as a lock of hair. To be fully appreciated the object has to be turned, caressed with the fingers, from trace to relic and back again, with that tactile experience of the relic so strongly linked to emotion.

What makes an object pleasurable to touch will clearly depend both on the macro- and microgeometric properties of the objects themselves as well as on the more cognitive/emotional associations that the particular object has to an individual – for example, the desire of some people to touch relics and other religious artefacts. It might seem that while the scientific approach to the study of touch may be able to help by providing guidelines concerning the particular micro- and macrogeometric properties of an object that will make it pleasant to touch, there is little that the scientific approach can say about the more emotional/cognitive (i.e. top-down) factors that modulate touch. However, this is not quite true.

For example, a colleague, Professor Bruce Hood in Bristol, has reported a nice example from his lectures on people's mystical beliefs where he describes handing a sweater to the audience when he gives a talk (personal communication, 9 November 2007). When he asks the audience how much they like to touch the sweater, people typically say that it feels fairly nice. But then when he tells the audience that the sweater once belonged to Fred West, the infamous serial killer (or to some other motley character), time and again, educated audiences will immediately discard the object, with some even refusing to touch it. This example clearly highlights the importance of top-down factors (e.g. cognition, beliefs, memory, etc.) in the tactile appreciation of objects (see also Nemeroff and Rozin 1994).

It would be fascinating to perform this kind of study in a brain scanner, perhaps using a method similar to that used so successfully by Rolls and his colleagues in their recent studies on the cognitive modulation of the pleasantness of interpersonal touch (see McCabe et al., 2008; Rolls, submitted). To put the experiment into a curatorially relevant context, one need only think of giving a group of participants a range of different stimuli and telling them that half the objects are antiques or ancient artefacts, while the other half are just regular household items (with the labels given to the participants counterbalanced across two different groups).[2] In our opinion, there is great scope for further research to help understand the psychological and neurocognitive principles underlying the foundations of a possible tactile 'aesthetics'. (On the difficulties associated with forming an aesthetics of touch, see Diaconu 2006, but also see Cranston 2003; Révész 1950; Hall 1999: 91).

Conclusions

Although a great deal of psychological, psychophysical and cognitive neuroscience research has investigated the sense of touch over the last century or so, very little of it, at least until very recently, has been able (or, for that matter, was designed) to address the problems associated with accessibility currently faced by museum and gallery curators. However, the last few years have seen the publication of a number of important papers that have started to move in the right direction. That is, there appears to be an emerging trend for researchers finally to move away from the assessment of sandpapers toward the evaluation of more realistic (and varied) ranges of experimental stimuli. Pairing this with the shift from the early focus of researchers on the sensory-discriminative aspects of touch towards the scientific assessment of the more aesthetic/affective attributes of objects, materials and surfaces, then one may be confident that there is scope for some real progress to be made in the next few years.

We see great research potential in the scientific approach to the study of the higher-level (i.e. more cognitive) factors affecting the tactile perception of both the sighted and the visually impaired (cf. Edwards 1999; McCabe et al. 2008; Nemeroff and Rozin 1994). We believe that it is only through further cooperation and constructive dialogue between the various different disciplines interested in the topic of enhancing accessibility that some tangible solutions to current problems faced by museum and gallery curators will be achieved. In the years to come, we hope to see researchers starting to study and develop tactile replicas designed to enhance the educational and aesthetic experiences of both the visually impaired and the sighted museum and gallery visitors,

Notes

1. Binns commented as follows: 'But the practical buyer of wool fabric seems unable, when looking at materials, to keep his hands away from them. He thinks and speaks in terms such as: pliable, supple, soft, kind, lofty, firm, hard, or harsh – and all these are associated with the sense of touch' (1926: 238).
2. In fact, Naomi Kendrick, an artist working in Liverpool, does just this in the sensory intervention art workshops that she organizes for people who are visually impaired (see http://www.filamentmedia.co.uk/elephant/contact.htm).

References

Arnold, K. (2003), 'Skulls, mummies and unicorns' horns: Medicinal chemistry in early English museums', in R.G.W. Anderson, M.L. Caygill, A.G. MacGegor and L. Syson (eds), *Enlightening the British: Knowledge, Discovery and the Museum in the Eighteenth Century*, London: British Museum Press, pp. 74–80.

Battaglia, P.W., Jacobs, R.A. and Aslin, R.N. (2003), 'Bayesian integration of visual and auditory signals for spatial localization', *Journal of the Optical Society of America A* 20: 1391–7.

Bergmann Tiest, W.M. and Kappers, A.M.L. (2006), 'Analysis of haptic perception of materials by multidimensional scaling and physical measurements of roughness and compressibility', *Acta Psychologica* 121: 1–20.

Bergmann Tiest, W.M. and Kappers, A.M.L. (2007), 'Haptic and visual perception of roughness', *Acta Psychologica* 124: 177–89.

Binns, H. (1926), 'The discrimination of wool fabrics by the sense of touch', *British Journal of Psychology* 16: 237–47.

Bolanowski, S.J., Verrillo, R.T. and McGlone, F. (1999), 'Passive, active and intra-active (self) touch', *Somatosensory and Motor Research,* 16, 304–11.

Candlin, F. (2003), 'Blindness, art and exclusion in museums and galleries', *Journal of Art and Design Education* 22: 100–10.

Candlin, F. (2004), 'Don't touch! Hands off! Art, blindness and the conservation of expertise', *Body and Society* 10: 71–90.

Candlin, F. (2006), 'The dubious inheritance of touch: Art history and museum access, *Journal of Visual Culture* 5: 137–54.

Classen, C. (2005), 'Touch in the museum', in C. Classen (ed.), *The Book of Touch*, Oxford: Berg, pp. 275–86.

Classen, C. (2007), 'Museum manners: The sensory life of the early museum', *Journal of Social History* 40: 895–914.

Classen, C. and Howes, D. (2006), 'The museum as sensescape: Western sensibilities and indigenous artefacts', in E. Edwards, C. Gosden and R.B. Phillips (eds), *Sensible Objects: Colonialism, Museums and Material Culture*, Oxford: Berg, pp. 199–222.

Cohen, L.G., Celnik, P., Pascual-Leone, A., Corwell, B., Faiz, L., Dambrosia, J., Honda, M., Sadato, N., Gerloff, C., Catala, M.D. and Hallett, M. (1997), 'Functional relevance of cross-modal plasticity in blind humans', *Nature* 389: 180–3.

Craig, J.C. (1977), 'Vibrotactile pattern perception: Extraordinary observers', *Science* 196: 450–2.

Cranston, J. (2003), 'The touch of the blind man: The phenomenology of vividness in Italian Renaissance art', in E.D. Harvey (ed.), *Sensible Flesh: On Touch in Early Modern Culture*, Philadelphia PA: University of Pennsylvania Press, pp. 224–42.

Demattè, M.L., Sanabria, D., Sugarman, R. and Spence, C. (2006), 'Cross-modal interactions between olfaction and touch', *Chemical Senses* 31: 291–300.

Diaconu, M. (2006), 'Reflections on an aesthetics of touch, smell and taste, *Contemporary Aesthetics,* downloaded from http://www.contempaesthetics.org/newvolume/pages/article.php?articleID=385, accessed 19 December 2007.

Edwards, E. (1999), 'Photographs as objects of memory', in M. Kwint, C. Breward and J. Aynsley (eds), *Material Memories: Design and Evocation*, Oxford: Berg, pp. 221–36.

Ember, M. (1978), 'Size of color lexicon: Interaction of cultural and biological factors', *American Anthropologist* 80: 364–7.

Ernst, M.O. and Banks, M.S. (2002), 'Humans integrate visual and haptic information in a statistically optimal fashion', *Nature* 415: 429–33.

Ernst, M.O. and Bülthoff, H.H. (2004), 'Merging the senses into a robust percept', *Trends in Cognitive Sciences* 8: 162–9.

Fiennes, C. (1949), *The Journeys of Celia Fiennes,* London: Cresset Press.

Ghazanfar, A.A. and Schroeder, C.E. (2006), 'Is neocortex essentially multisensory?', *Trends in Cognitive Sciences* 10: 278–85.

Guest, S., Catmur, C., Lloyd, D. and Spence, C. (2002), 'Audiotactile interactions in roughness perception', *Experimental Brain Research* 146: 161–71.

Guest, S. and Spence, C. (2003a), 'Tactile dominance in speeded discrimination of pilled fabric samples', *Experimental Brain Research* 150: 201–7.

Guest. S. and Spence, C. (2003b), 'What role does multisensory integration play in the visuotactile perception of texture?', *International Journal of Psychophysiology* 50: 63–80.

Guest, S. and Spence, C. (2007), 'Tactile-to-visual matching of pilled fabric samples, *submitted for publication.*

Hall, J. (1999), *'The world as sculpture',* London: Random House.

Heller, M.A. (1983), 'Haptic dominance in form perception with blurred vision', *Perception* 12: 607–13.

Heller, M.A. (1989), 'Texture perception in sighted and blind observers', *Perception and Psychophysics* 45: 49–54.

Heller, M.A. (2004), 'Mind and body: Psychology and neuroscience', *Perception* 33: 383–5.

Hollins, M. (1989), '*Understanding Blindness*, Hillsdale NJ: Erlbaum.

Hollins, M., Bensmaïa, S., Karlof, K. and Young, F. (2000), 'Individual differences in perceptual space for tactile textures: Evidence from multidimensional scaling', *Perception and Psychophysics* 62: 1534–44.

Hollins, M., Faldowski, R., Rao, S. and Young, F. (1993), 'Perceptual dimensions of tactile surface texture: A multidimensional scaling analysis', *Perception and Psychophysics* 54: 697–705.

Hooke, R. (1971[1705]), *The Posthumous Works of Robert Hooke*, R. Waller (ed.), 2nd facsimile edn, London: Smith and Walford.

Iggo, A. (1977), 'Cutaneous and subcutaneous sense organs', *British Medical Bulletin* 33: 97–102.

Klatzky, R.L., Lederman, S.J. and Matula, D.E. (1993), 'Haptic exploration in the presence of vision', *Journal of Experimental Psychology: Human Perception and Performance* 19: 726–43.

Klatzky, R.L., Lederman, S.J. and Metzger, V.A. (1985), 'Identifying objects by touch: An "expert system"', *Perception and Psychophysics* 37: 299–302.

Klatzky, R.L., Lederman, S. and Reed, C. (1989), 'Haptic integration of object properties: Texture, hardness, and planar contour', *Journal of Experimental Psychology: Human Perception and Performance* 15: 45–57.

Kóbor, I., Füredi, L., Kovács, G., Spence, C., and Vidnyánszky, Z. (2006), 'Back-to-front: Improved tactile discrimination performance in the space you can't see', *Neuroscience Letters* 400: 163–7.

Laird, D.A. (1932), 'How the consumer estimates quality by subconscious sensory impressions: With special reference to the role of smell', *Journal of Applied Psychology* 16: 241–6.

Lederman, S.J. (1982), 'The perception of texture by touch', in W. Schiff and E. Foulke (eds), *Tactual Perception*, Cambridge: Cambridge University Press, pp. 130–67.

Lederman, S.J. and Abbott, S.G. (1981), 'Texture perception: Studies of intersensory organisation using a discrepancy paradigm and visual versus tactual psychophysics', *Journal of Experimental Psychology: Human Perception and Performance* 7: 902–15.

Lederman, S. J. and Klatzky, R. L. (1990), 'Haptic classification of common objects: Knowledge-driven exploration', *Cognitive Psychology*, 22, 421–59.

Lederman, S.J. and Klatzky, R.L. (2004), 'Multisensory texture perception', in G.A. Calvert, C. Spence and B.E. Stein (eds), *The Handbook of Multisensory Processes*, Cambridge, MA: MIT Press, pp. 107–22.

Lederman. S.J., Thorne, G. and Jones, B. (1986), 'Perception of texture by vision and touch: Multidimensionality and intersensory integration', *Journal of Experimental Psychology: Human Perception and Performance* 12: 169–80.

Losche, D. (2006), 'The fate of the senses in ethnographic modernity: The Margaret Mead Hall of Pacific peoples at the American museum of natural history', in E. Edwards, C. Gosden and R.B. Phillips (eds), *Sensible Objects: Colonialism, Museums and Material Culture*, Oxford: Berg, pp. 223–44.

MacGregor, G. (1999), 'Making sense of the past in the present: A sensory analysis of carved stone balls', *World Archaeology* 31: 258–71.

Maffei, L. and Fiorentini, A. (1995), *Arte e cervello* [Art and the brain], Bologna: Zanichelli.

Major, D.R. (1895), 'On the affective tone of simple sense-impressions', *American Journal of Psychology* 7: 57–72.

Marinetti, F.T. (1921), *Il manifesto del tattilismo* [The Futurist manifesto], Milan: Comoaedia.

McCabe, C., Rolls, E.T., Bilderbeck, A. and McGlone, F. (2008), 'Cognitive influences on the representation of touch in the human brain', *Social, Cognitive, and Affective Neuroscience* 3: 97–108.

McGlone, F., Vallbo, A.B., Loken, L. and Wessberg, J. (2007), 'Discriminative touch and emotional touch', *Canadian Journal of Experimental Psychology* 61: 173–83.

McLoone, H.E. (2003), 'Touchable objects: Attributes applied to the design of computer input devices', *Ergonomics* 46: 1320–31.

Merabet, L., Thut, G., Murray, B., Andrews, J., Hsiao, S. and Pascual-Leone, A. (2004), 'Feeling by sight or seeing by touch?', *Neuron* 42: 173–9.

Minogue, J. and Jones, M.G. (2006), 'Haptics in education: Exploring an untapped sensory modality', *Review of Educational Research* 76: 317–48.

Montessori, M. (1991), *Educazione per un mondo nuovo* [Education for a new world], Milan: Garzanti.

Nemeroff, C. and Rozin, P. (1994), 'The contagion concept in adult thinking in the United States: Transmission of germs and interpersonal influence', *Ethos* 22: 158–86.

Ouzman, S. (2001), 'Seeing is deceiving: Rock art and the non-visual', *World Archaeology* 33: 237–56.

Picard, D., Dacremont, C., Valentin, D. and Giboreau, A. (2003), 'Perceptual dimensions of tactile textures', *Acta Psychologica* 114: 165–84.

Power, R.P. (1981), 'The dominance of touch by vision: Occurs with familiar objects', *Perception* 10: 29–33.

Power, R.P. and Graham, A. (1976), 'Dominance of touch by vision: Generalization of the hypothesis to a tactually experienced population', *Perception* 5: 161–6.

Putzar, L., Goerendt, I., Lange, K., Rösler, F. and Röder, B. (2007), 'Early visual deprivation impairs multisensory interactions in humans', *Nature Neuroscience* 10: 1243–5.

Révész, G. (1950), *'The psychology and art of the blind'*, London: Longmans Green.

Röder, B., Rösler, F. and Spence, C. (2004), 'Early vision impairs tactile perception in the blind', *Current Biology* 14: 121–4.

Rolls, E.T. (in press), 'The cognitive and affective processing of touch in the brain', *Neuroscience and Biobehavioural Reviews*.

Rolls, E.T., O'Doherty, J., Kringelbach, M.L., Francis, S., Bowtell, R. and McGlone, F. (2003), 'Representations of pleasant and painful touch in the human orbitofrontal and cingulate cortices', *Cerebral Cortex* 13: 308–17.

Sadato, N., Pascual-Leone, A., Grafman, J., Ibañez, V., Deiber, M.-P., Dold, G. and Hallett, M. (1996), 'Activation of the primary visual cortex by Braille reading in blind subjects', *Nature* 380: 526–8.

Saito, D.N., Okada, T., Honda, M., Yonekura, Y. and Sadato, N. (2007), 'Practice makes perfect: The neural substrates of tactile discrimination by Mah-jong experts include the primary visual cortex', *BMC Neuroscience* 7: 79.

Schifferstein, H.N.J. and Spence, C. (2007), 'Multisensory product experience', in H.N.J. Schifferstein and P Hekkert (eds), *Product Experience*, Oxford: Elsevier, pp. 133–61.

Silvera, D.H., Josephs, R.A. and Giesler, R.B. (2002), 'Bigger is better: The influence of physical size on aesthetic preference judgments', *Journal of Behavioural Decision Making* 15: 189–202.

Spence, C. (2008), 'Making sense of touch: A multisensory approach to the perception of objects', in E. Pye (ed.), *The Power of Touch: Handling Objects in Museums and Heritage Contexts',* Walnut Creek, CA: Left Coast Press, pp. 45–61.

Spence, C. and Zampini, M. (2006), 'Auditory contributions to multisensory product perception', *Acta Acustica united with Acustica* 92: 1009–25.

Suganami, M. (2003), 'Experiencing Art', paper presented at Translating Art for the Blind Seminar, 6 June, Tate Modern, London.

Warren, D.H. and Rossano, M.J. (1991), 'Intermodality relations: Vision and touch', in M.A. Heller and W. Schiff (eds), *The Psychology of Touch*, Hillsdale, NJ: Lawrence Erlbaum, pp. 119–37.

Weinstein, S. (1968), 'Intensive and extensive aspects of tactile sensitivity as a function of body part, sex, and laterality', in D.R. Kenshalo (ed.), *The Skin Senses*, Springfield, IL: Thomas, pp. 195–222.

Yoshida, M. (1968a), 'A psychometric analysis of textile handling', *Japanese Psychological Research* 10: 1–12.

Yoshida, M. (1968b), 'Dimensions of tactual impressions (1)', *Japanese Psychological Research* 10: 123–37.

Yoshida, M. (1968c), 'Dimensions of tactual impressions (2)', *Japanese Psychological Research* 10: 153–73.

Zampini, M., Mawhinney, S. and Spence, C. (2006), 'Tactile perception of the roughness of the end of a tool: What role does tool handle roughness play?', *Neuroscience Letters* 400: 235–9.

Zangaladze, A., Epstein, C.M., Grafton, S.T. and Sathian, K. (1999), 'Involvement of visual cortex in tactile discrimination of orientation', *Nature* 401: 587–90.

–3–

The Two Sides of Touch: Sensing and Feeling

Francis McGlone

The skin senses have traditionally been described as telling us (our brains) about the nature of tactile, thermal or painful stimulation that impinge on the body surface, and we have a relatively robust understanding of their peripheral and central neural counterparts – collectively known as the *somatosensory system.* It is becoming clear that the sense of touch serves much more than pure discrimination – for example, telling us how hard, cold or smooth something is – as it also conveys information about how pleasant or rewarding – as well as unpleasant – the experience of touch can be, such as when being touched by somebody or some thing. There is now emerging evidence that there is an emotional aspect to touch – called affective touch – as well as a discriminative one. But very little is known about how tactile experiences such as those sensed from the hand, such as when handling objects in a heritage context, contrast with the touch we experience from the rest of our body surface where, research is discovering, we are more sensitive to the emotional rather than the discriminative aspects of touch.

It has been recognized for some time that in all mammals there is a class of peripheral nerve fibres (called c-fibres) in the skin of the body (not in the skin of the hand) that respond to light mechanical stimulation (stroking), but that in humans this system had been viewed as vestigial. C-fibres are traditionally classified as either subserving pain sensibility or autonomic skin (e.g. control of blood vessels or sweat glands), but there is increasing evidence that a functional 'pleasure' c-fibre system exists in human skin, notably found only in hairy skin – that is, all skin sites other than the palms of the hands or the soles of the feet. Employing a direct nerve recording technique (microneurography), researchers have found a population of c-fibres in humans that are preferentially activated by gentle stroking movements and not, as is traditionally found with c-fibres, by painful stimuli. There is also recent evidence, employing behavioural as well as brain-imaging techniques, of how the body responds to being stroked by different textures and where in the brain the 'pleasure' nerves project their messages.

Introduction

Sensory systems are translators that throughout evolution have enabled all life forms, from the single-celled organisms such as amoeba and paramecium through to humans, to interpret their physical and chemical environment – to sense photons, to detect minuscule vibrations, to analyse airborne volatile chemicals, to identify essential chemical fuel sources. However, if simply sensing and interpreting the immediate environment was all the senses did then life would not have evolved very far. A seminal event occurred many millions of years ago that led to a divergence in the paths the evolution of living organisms would take, rendering one form essentially trapped in endless time, while providing the other with a road to world domination. The sponges and the anemones are the two protagonists in this saga, and the simple/single event that led to their separate destinies was coupling *sensing* with *moving* through the evolution of contractile cells. The sponge never did move, so is still a sponge; the anemone did, and is now us. By being able to explore its environment with its tentacles, the anemone was no longer dependent on food finding it, but was able to actively search for it – and the rest is evolutionary history.

This primal drive to explore manifests itself in human terms as a characteristic we call 'motivation' – the desire to get off the couch and seek stimulation. Although our interactions with the environment are essentially multisensory ones (as Charles Spence describes in this volume), our knowledge of the brain processes driving and controlling these behaviours has been largely understood by psychologists and neuroscientists in terms of just two out of the five sensory modalities – vision and hearing. These senses are classified as exteroceptive, providing information to the brain that can be used to guide approach or avoidance behaviours, and are often associated with emotional states such as reward and aversion. One other sense, olfaction, can also provide such distant information, such as the aroma of fresh coffee and the smell of burning. But on many occasions, just looking, hearing or smelling is not enough and can even lead to negative emotional states such as frustration. In these situations we require the deployment of another sense in order to enhance sensory information about objects in the environment – the sense of touch (Mountcastle 1984; Katz 1989[1925]; Darian-Smith 1984). We rely on this sense to manipulate and explore the physical properties of objects and, via haptic processes, identify their shape and function. Without the ability to detect mechanical forces (touch) it would be impossible to use tools or, as we shall see later, even to move. In a museum context there is a problem with allowing visitors access to exhibits through senses other than sight, and particularly that of touch – a sense that is critically important for the visually impaired visitor for example. Museums are concerned of course that the touching and handling of their artefacts creates risk to their conservation but as Hetherington (2000, 2002, 2003) has pointed out in his research, the opportunity to touch objects (the Tate Gallery first did this in a pioneering exhibition in 1976) is highly rewarding.

What has just been described is known as 'active touch' – touch perceived as a consequence of movement - but there is a second form of touch, known as 'passive touch', where movement of the self is not required and which is normally experienced when another person or object is touching you. This form of touch plays a very important role in social communication behaviours involving inter-personal touch (one person touching another) such as grooming or nurturing. Here, touch can communicate so much more than language – such as the hand on the shoulder in a time of need, the massage after a hard day's work, the stroking of a crying baby (Hertenstein et al. 2006). In all of these instances, it is the toucher's hand that is delivering the received touch, and in particular the palmar surface of the hand. We don't stroke a baby with the backs of our hands, but we will often explore textures and surfaces that way (see later in the chapter).

There is one final form of touch, called intra-personal touch (touching oneself), that needs to be described in order to understand the consequences of touching one's body either with one's own hand, or with an object grasped by the hand, such as a comb or an artefact. In these situations the body surface being touched is experiencing 'passive touch' whereas the body part delivering the touch – the hand – is involved in 'active touch'. Recent neuroscientific research is providing better understanding of how these various forms of touch differ and why. With object handling, for example, the touched rather than the touching body part may communicate more about the emotional attributes of the object rather than its purely physical attributes. 'Touch' in the contexts just described is operating as an interoceptive sensory modality, providing information to the brain about the state of the body surface and that of close encounters with the world around, be they animate or inanimate. These states are also often associated with emotional consequences such as reward and aversion.

Exploratory behaviours involve all of the senses and we now know that the brain utilizes information from the individual senses to enhance perception, via mechanisms known as multisensory integration, whereby inputs from the primary modalities (vision, touch, etc.) are pooled to extract the most amount of information about events or objects in the environment. However, the primary sensory modality serving the body senses, commonly described as *touch*, is itself multisensory. It is collectively described as the *somatosensory system*, and comprises all those peripheral nerve fibres and specialized receptors that serve not only cutaneous sensitivity (the sensory receptors found in the skin) but also proprioception (sensory receptors found in the joints and muscles). Thus it provides information about limb position and muscle forces which the central nervous system uses to monitor and control limb movements and ensure that actions and movements are executed fluently.

The cutaneous senses provide the brain with information from a range of multimodal sensory receptors and nerves in the skin and comprise the four main submodalities of touch, temperature, itch and pain (Willis 1985; Darian-Smith 1984) However, there is growing evidence for what is considered to be a fifth submodality, the nerve receptors of which are preferentially activated by slowly moving, low force, mechanical stimuli and are posited to play a role in the affiliative or emotional (pleasant) aspects of touch. A key differentiating factor in this latter system is that it is not present in the palmar skin of the hand – so-called glabrous skin – but found only in the hairy skin of the body – essentially all skin outside the palms and soles of the feet is described as 'hairy skin'.

Exploratory tactile behaviours cover a wide range of interactions with the physical world from searching for a lost contact lens to handling a Ming vase. Of interest here is the role of the two tactile systems outlined above when touching and handling objects in a cultural heritage context, where touch can serve discriminative (is it hot, rough, hard, etc.?), haptic (its geometric and categorical properties), as well as affective or emotional functions (is it pleasant, unpleasant to feel?). Recent neuroscientific research in somatosensation could lead us to an understanding of how museum curators could better select objects for touching and handling that would provide a range of sensations and emotions, from pleasant to strange, or even unpleasant. This, in turn, could help guide visitors to use their innate exploratory tactile behaviours in novel ways – by employing different parts of the body, such as the backs of the hands, or to use a range of different exploratory movements to achieve different sensory effects.

These two sides of touch – the discriminative and the affective – will now be described and, based on recent findings on the role of the 'fifth' cutaneous sensory sub-modality in translating the pleasant aspects of touch, new ways in which interaction with heritage objects might be discovered.

Discriminative Touch

Research into skin sensory processing has largely focused on the glabrous surface of the hand, in particular the digits (Johansson 1976; Darian-Smith 1984; Gescheider et al. 1992; Greenspan and Lamotte 1993; Vallbo et al. 1979; Willis and Coggeshall 1991). Of the four 'classical' cutaneous submodalities of the somatosensory system, touch subserves the perception of pressure, vibration, slip and texture, and is critical in providing haptic information about handled objects. It relies upon four different receptors in the digit skin: (1) Meissner's corpuscles, (2) Pacinian corpuscles, (3) Merkel's disks, and (4) Ruffini endings, collectively known as low threshold mechanoreceptors (LTMs), a class of receptors specialized to transduce (or 'translate') the mechanical forces impinging the skin into the nerve impulses (called spikes) that convey the information towards the central nervous system in the form of binary code. In handling objects it is vital that the tactile information is conveyed as quickly as possible to the brain as this information is used 'on-line' to guide and control the hand muscles. If this were not the case, then controlling how much force to apply in handling a delicate object or detecting the object slipping in the hand would not be possible.

To enable this information to get to the brain quickly, LTMs are connected to nerve fibres that are specialized to transmit spikes at up to 100 metres per second, ensuring the tactile information gets to brain within tens of milliseconds. Sensory nerves contain thousands of individual nerve fibres, called axons, and these are classified according to their degree of myelination. Myelin is a fatty sheath that surrounds the nerve fibre rather like the insulation around an electrical cable, and the amount of myelination (how thick it is) determines the speed with which the axon can conduct nerve impulses and hence the nerve's conduction velocity (Table 3.1).

Table 3.1 Main characteristics of primary sensory afferents innervating human skin

	Sensory Afferent Nerves		
Class	*Modality*	*Axonal diameter (μm)*	*Conduction velocity (m/sec)*
Myelinated			
A-beta	Low threshold mechanoreceptors	10	80
A-δ	Cold, noxious, thermal	2.5	12
Unmyelinated			
C-Pain	Noxious, heat, thermal	1	<2
C-Itch	Itch	1	<1
C-Tactile	Light stroking, gentle touch	1	<2
C-Autonomic	Autonomic, sweat glands, vasculature	1	<2

Primary sensory afferents innervating human skin are either myelinated or unmyelinated and, as can be seen in Table 3.1, the presence or absence of a myelin sheath affects the nerves conduction velocity, with the unmyelinated c-fibres conducting nerve impulses the slowest.

The largest and fastest axons are called A-alpha and include some of the proprioceptive neurons, such as the muscle-stretch receptors. The second largest group, called A-beta, includes all of the discriminative touch receptors being described here. Pain, itch and temperature include the third and fourth groups, A-δ and C fibres. The Meissner and Pacinian receptors are classified as fast adapting (FA) and respond best to a temporally or spatially moving mechanical stimulus on the skin, and the Merkel and Ruffini receptors are classified as slowly adapting (SA) and generate spikes during a constant mechanical stimulus. A further classification relates to the LTMs receptive field (RF) – that is, the surface area of skin to which they are sensitive. The RF is determined by the LTMs anatomical location within the skin, with those near the surface at the dermal/epidermal boundary, Meissner's corpuscles and Merkel's disks, having small RFs, and those lying deeper within the dermis, Pacinian corpuscles and Ruffini endings, having large RFs. LTMs can be

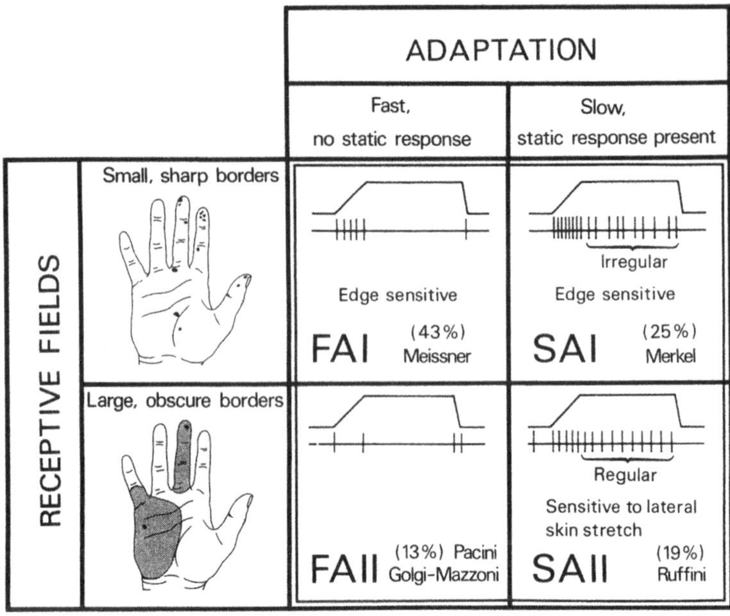

Figure 3.1 The four types of low threshold mechanoreceptors in human glabrous skin (after Westling 1986).

seen as tiny microphones in the skin that 'listen' to mechanical events happening on the surface of the skin (Figure 3.1). The four panels (FAI, SAI, FAII, SAII) in the centre of Figure 3.1 show the nerve-firing responses to a ramp and hold indentation, and the percentage frequency of occurrence and putative morphological correlate. The black dots in the left panel show the receptive fields of Type I (top) and Type II (bottom) afferents. The right panel shows the average density of Type I (top) and Type II (bottom) afferents with the darker area depicting higher densities.

Researchers have described four distinct channels mediating tactile perception in the glabrous skin of the hand (Bolanowski et al. 1988), with each channel being represented by one of the four LTMs. Psychophysicists study the sense of touch by using vibratory stimuli delivered to the surface of the skin by electromechanical shakers and have found that frequencies in the 40–500 Hz range generate a sense of 'vibration', and that Pacinian corpuscles were maximally active, whereas Meissner corpuscles are responsible for the sense of 'flutter' experienced in the 2–40 Hz range. The sense of 'pressure' was mediated by Merkel's disks in the 0.4–2.0 Hz range, and Ruffini end organs responding in the 100–500 Hz range played a role in sensing skin stretch (Table 3.2).

In the late 1960s two Swedish neuroscientists developed a direct nerve recording technique that revolutionized our understanding of the sense of touch in humans.

Table 3.2 Frequency responses and types of sensations experienced by electrical microstimulation of low threshold mechanoreceptors found in the human hand

Channel	*FAII*	*FAI*	*SAII*	*SAI*
Frequency response	40–800 Hz	2–40 Hz	100–500 Hz	<0.3 to >100 Hz
Sensation	Vibration	Flutter	Not known	Pressure
Receptor type	FAII Pacinian corpuscle	FAI Meissner's corpuscle	SAII Ruffini end organ	SAI Merkel's disk

Note: There is no description of a sensation to electrical microstimulation for SAII receptors as none has been found so far.

The technique was developed by Hagbarth and Vallbo (1967) and is called microneurography. It involves inserting a fine tungsten microelectrode with a tip diameter of less than 5 μm through the skin and into an underlying peripheral nerve. For studies on the glabrous skin of the hand the microelectrode is inserted at the wrist, where one of the main nerves innervating the hand is found. This nerve, the median nerve, contains thousands of individual nerve fibres (called axons) each of which connects with one or other of the LTMs in the hand. The technique is akin to tapping into a telephone line and listening to a conversation. A sensitive biological amplifier records and amplifies the spike discharges conveyed by the axons and feeds these to a loudspeaker to enable the experimenter to hear the spike activity and 'home-in' on a single nerve fibre. Skilled manual micromanipulation of the electrode, coupled with stroking across the hand to stimulate LTMs, results first in a population response being recorded – that is, neural activity in a nerve bundle containing hundreds of axons – until finally, sometimes after many hours of 'fishing', a single one is isolated (a peripheral nerve can be likened to an electric cable containing thousands of wires). At this stage the receptive field of the single nerve fibre is mapped using a fine nylon filament (von-Frey hair) and the unit subtype (i.e. FA or SA) is identified. Once this stage is complete a small pulsed current of a few microamps (typically less than 10 μA) is delivered through the electrode that electrically stimulates the nerve and provides a perceptual confirmation of the unit subtype. If, for example, an FA unit has been recorded, microstimulation is perceived as a 'flutter' or 'vibration', depending on the frequency of the electrical stimulation pulses, and is perceptually localized to the previously mapped receptive field – that is, the subject experiences 'touch' as coming from the precise part of the skin where the RF was mapped. Figure 3.1 depicts the relationships between RF, adaptation rate and unit type from studies carried out on the human hand (Westling 1986). Electrophysiological studies by Vallbo and Johansson (1978) on single peripheral nerve fibres innervating the human hand have provided a generally accepted model of touch that relates the four anatomically defined types of LTMs to their neural (spike) response patterns.

The concentration on the hand with researches into touch has led to relatively few studies of tactile sensitivity being carried out on hairy skin. Johansson et al. (1988) and Essick and Edin (1995) have described sensory fibres with mechanical properties in human hairy skin, but the relationship between these sensory fibres and tactile perception is still uncertain, as exemplified by the response properties of a class of slowly adapting touch nerves (SAI afferents) that Harringhton and Merzenich (1970) found to be responsive to levels of mechanical stimulation that were below perceptual thresholds. Jarvilehto et al. (1976) also described high levels of electrical activity in these nerves that was not perceivable, in contrast to the responses of this same class of afferents in glabrous skin where SAI nerve activity is directly correlated with a conscious sense of pressure. These observations may be relevant when we come to review the role of a class of emotional touch nerves called c-tactile afferents (CT-afferents).

Sensations from the skin, encoded in peripheral nerves as spike activity, enter the central nervous system via the dorsal half of the spinal cord and form nerve fibre bundles (tracts) that ascend to the top of the spinal cord. Here they form connections (called synapses) with second-order neurones that relay the information, after crossing over to the other side of the brain, to a major sensory junction-box in the centre of the brain called the thalamus. From here, third order neurones project to the overlying primary and secondary somatosensory cortex, where the contralateral body surfaces are mapped in each hemisphere – that is, the left side of the body projects to the right side of the brain and vice versa. The sensory information from all body surfaces is mapped in a somatotopic (body-mapped) manner (Penfield and Rasmussen 1952; Maldjian et al. 1999), with the legs represented medially, at the top of the head, and the face represented laterally. In line with other sensory modalities, information is then passed on to higher-order brain structures that control such cognitive processes as attention, memory and emotion, as well as systems that integrate this information with information coming from the other sensory modalities that enable the brain to maximize the information it receives from the senses about conditions in the external world – multisensory perception.

Within the cortex of the brain there are thought to be at least four separate areas processing somatosensation – the touch brain – (i) the primary somatosensory cortex (SI); (ii) the secondary somatosensory cortex (SII) (Woolsey 1946; Maeda 1999; Coghill et al. 1994; Francis et al. 2000; McGlone et al. 2002); (iii) the insular cortex (Schneider et al. 1993); (iv) the posterior parietal cortex (Mesulam 2000).

As with studies of the peripheral nervous system, the technique of micro-neurography has again been employed in studies of the central nervous somatosensory system – in this case, to study the relationship between finger sensory nerves and their central projections in the brain – by combining it with a brain imaging technique called functional magnetic resonance imaging (fMRI) that allows us to 'see' which parts of the brain are activated when a part of the body is touched or a tactile sensory nerve is electrically stimulated. (The brain cannot tell the difference between an 'electric' touch and a normal mechanical touch.) We have found that electrical microstimulation of individual LTM afferents in the median nerve, innervating the fingers of the hand, produces activations in the primary and secondary somatosensory cortices (Trulsson et al. 2000). This is in accordance with the findings of Penfield and Boldrey (1937), and is allowing researchers to study the human 'hand touch brain' in much greater detail. All of this research is focused on the sense of touch from the palm of the hand and is key to our understanding of that form of touch experienced when we handle tools or objects. However, as we have alluded to above, there is another form of touch – namely, emotional touch.

Emotional Touch

As already described, human discriminative tactile sensibility is mediated by LTMs with fast-conducting, large myelinated afferents, but in recent years a growing body of evidence has been accumulating from anatomical, psychophysical, behavioural, electrophysiological and neuroimaging studies that a further submodality of afferent slowly conducting, but unmyelinated, nerves exist in human hairy skin. These are neither pain- nor itch-sensitive, but respond preferentially to low force, slowly moving mechanical stimuli, and that might be coding for the affective or pleasant properties of touch. These nerve fibres have been classified as C-tactile afferents (CT-afferents) and were first reported in humans by Johansson et al. (1988), and subsequently by Nordin (1990), employing the technique of microneurography outlined above. These nerve types have never been found in glabrous skin (Vallbo et al. 1979; Vallbo et al. 1993; Edin 2001), and it is here we have our first clue as to their special properties, as we shall see later. It is well known that low-threshold mechanoreceptive innervation of the skin of many mammals is subserved by both A and C afferents (Zottermann 1939; Bessou and Perl 1969; Iggo and Kornhuber 1977), but until the observations of Johansson et al. (1988), Nordin (1990) and Vallbo et al. (1973), mechanosensitive C-afferents in human skin appeared to be entirely lacking. Figure 3.2a shows the experimental set-up employed for microneurographic recording from CT-afferents, and it can be seen that the electrode enters the skin through the upper arm, rather than at the wrist, when glabrous skin LTMs are recorded, entering a peripheral nerve that innervates the hairy skin on the back of the hand.

In order to study CT-afferents a robotic stimulator was developed that provided a high degree of programmable control over both the speed and force with which a surface was stroked across the skin (see Figure 3.2b). This device is called the Rotary Tactile Stimulator (RTS), and using this in microneurography experiments we have measured the electrophysiological and psychophysical properties of CT-afferents and been able to compare them with recordings from myelinated LTMs.

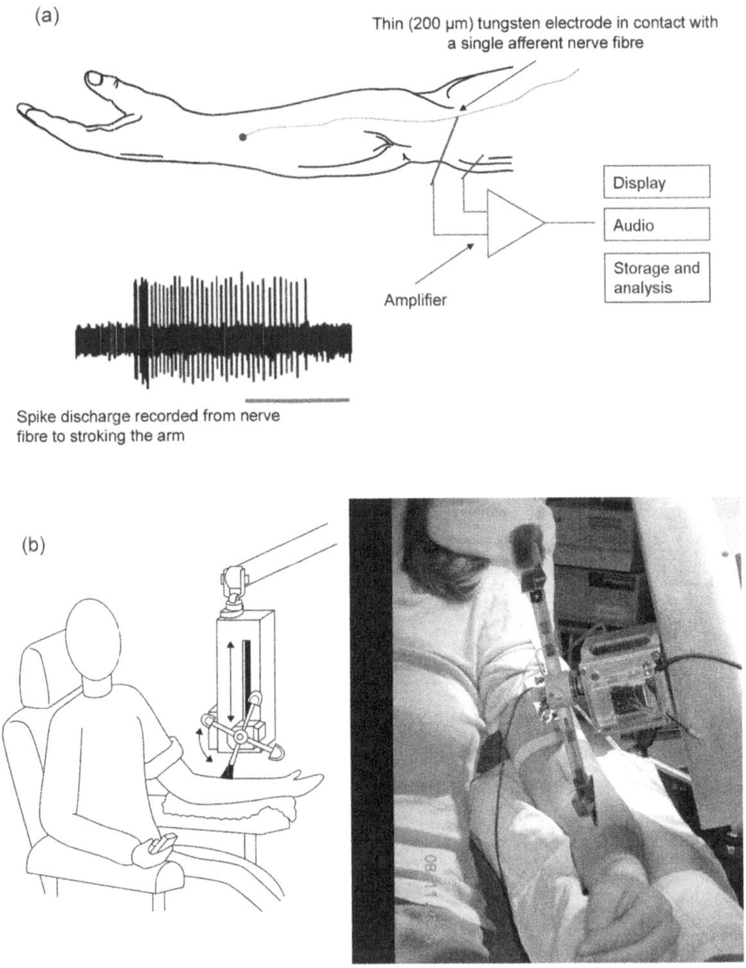

Figure 3.2 (a) The electrophysiological set-up for recording from CT-afferents. *Note*: Nerve impulses are recorded in response to stroking across the unit's receptive field either using a soft brush or the RTS. (b) A purpose built rotary tactile stimulator (RTS). *Note*: The RTS is shown in diagrammatic form on the left with an actual image of it in use during a nerve recording experiment on the right. The RTS allows the delivery of controlled force and velocity stimulation to the receptive fields of a CT unit and fast and slowly adapting LTMs.

One interesting observation was the stroking-velocity dependence of CT-afferents compared with LTMs. SAs and RAs increased their spike frequency to increasing stimulus velocities, whereas CTs were 'tuned', displaying a peak spike impulse frequency at stroking velocities in the range 1–5 centimetres per second, the spike discharge activity dropping off either side of this frequency. This 'preferred' stroking speed can be observed in human affiliative interactions, such as when a mother strokes her baby. Although CTs have been shown to activate more strongly to moving tactile stimulation, they are also activated by static touch as when a hand or object is gently placed on the skin.

By comparing results obtained from psychophysical experiments using the RTS with data from microneurography experiments, it has been shown (Loken et al. 2006) that the same stroking stimulus velocities that are reported by subjects as being the most pleasant are precisely the same as the optimal responses recorded from single-unit spike discharge frequencies of CTs, namely 1–5 centimetres per second. This concordance of neurophysiological and psychophysical results provides further evidence for the role of CT-afferents in emotional touch. Although we are still not clear as to the functional role of CT-afferents in somatosensation, their electrophysiological response properties, nerve fibre class and slow conduction velocities preclude their role in any rapid mechanical discriminative or cognitive tactile tasks and point to a more emotional (limbic) function (Vallbo et al. 1993; Essick et al. 1999).

So far we have described the psychophysical and electrophysiological properties of CTs, but in order to find out where in the skin CT-afferents were anatomically located – in other words, are they in the dermis with the other LTMs or out in the epidermis? – Anderson et al. (1999) and Reilly et al. (1997), employing a nerve fibre specific histological staining technique identified a population of very small nerve endings located solely within the epidermis that were hypothesized to be CT-afferents. Support for this hypothesis has recently been provided in a study by Liu et al. (2007) that employed a genetically encoded tracer (in mice) that has found a subpopulation of unmyelinated sensory nerve fibres which exclusively innervate hairy skin and which terminate in large receptive fields similar in size and distribution to CT-afferents described in humans by Wessburg et al. (2003).

As with peripheral nerve input to the central nervous system in the discriminative tactile system, CTs enter via the dorsal half of the spinal cord, but unlike A-betas, do not immediately ascend towards the brain but synapse in the dorsal horn of the spinal cord, where second-order fibres then cross over to the contralateral side and ascend in a fibre-bundle called the spinothalamic tract. Craig (2002) describes onward projections in this tract to two thalamic sites that activate a part of the emotional brain – called the limbic system – and another more sensory part. According to Craig (2003) these two projections correlate with the sensory discriminative and the emotional motivational aspects of feelings derived from the body. Both are strongly interconnected with the brain structures that are known to be involved in emotion, such as the amygdala, hypothalamus and orbitofrontal cortex. Recent neuroimaging studies have provided convergent data confirming the role of a key region of the cortex – the dorsal posterior insula – as one of the primary cortical targets for not only pleasant touch but also pain, temperature and itch (Francis et al. 1999; Rolls et al. 2003).

Two brain imaging techniques have proved extremely useful in helping us understand how these two forms of touch – discriminative and emotional – are represented in the human brain. Magnetic resonance imaging (MRI), developed in the 1980s, uses very powerful magnets to detect signals from protons – particles with a positive electronic charge that act like compass needles in the magnetic field and which are found in abundance in the body in water molecules. In the early 1990s physicists found ways to adapt MRI, which was being used to define brain structures, to also measure functional changes in brain activity. This finding relied upon a fortuitous property of blood where changes in the amount of oxygen it contains affects its magnetic properties. With functional MRI (fMRI), regions in the brain where changes in the levels of blood oxygenation due to neuronal activity-related changes in blood flow are detected, provide both anatomical and functional information enabling researchers to determine which brain regions are active in a sensory, emotional or cognitive task. The second brain imaging technique – predating fMRI – is called positron emission tomography (PET). By injecting small amounts of radioactive substances, or isotopes, into the blood, these isotopes release particles known as positrons, which produce other particles called photons that can be detected by special cameras. When radioactive water, labelled with the isotope oxygen 15 (O^{15}), is injected into the blood, it is taken into the brain in proportion to increased blood flow and acts as a measure of nerve cell, or neuron, activity in different brain areas. Using PET and O^{15} labelled water it is possible to locate regions in the brain that become active while a person sees or thinks about solving a problem, and by comparing images taken during a particular task with those taken while the brain is resting it is possible to expose brain structures involved in that task.

Employing these neuroimaging techniques we have been able to find out where in the brain CTs project, plus some clues as to their functional role. By working with two subjects who suffer from a rare neurological condition in which all their A-fibres (touch) have been destroyed, we have been able to dissociate 'normal' discriminative touch from affective or emotional touch. One consequence of this condition is that it renders sufferers not only completely insensitive to touch but also unable to move because the mechanosensitive proprioceptive nerves, which tell the brain where a limb is in space and how much force a muscle is applying to a load, are also destroyed. Without this information movement is normally impossible. However, one of the subjects has literally 'willed' himself to move (see Cole 1995). When these subjects were gently stroked on a hairy skin site, it was discovered that this produced a faint sensation of pleasant touch even though they had no experience of being touched, and when their brains were scanned with fMRI, activation was found in the posterior insular cortex but not in the primary somatosensory cortex. We interpret this as evidence that CT-afferents provide a system for limbic touch that might underlie emotional, hormonal and affiliative responses to skin-to-skin contacts between individuals engaged in bonding and nurturing behaviours (Olausson et al.

2002; Wessberg et al. 2003; McGlone et al. 2007). When touched on glabrous skin sites, the subjects were unable to report any sensations and when their brain scans were looked at there was no activity in either primary sensory cortex or insular cortex. It is possible that pleasant touch stimulates opioid receptors on these peripheral nerves and that this signal is decoded in emotional processing brain areas as has been described by Pecina et al. (2006). Further evidence of the representation of pleasant touch in the brain, this time using neurologically normal subjects, has been provided by Francis et al. (1999), where it was shown that the discriminative and emotional aspects of touch are processed in different brain areas. Activation of the primary somatosensory cortex was found to the physical aspects of stimulation, whereas the orbitofrontal cortex (an area of the frontal lobes involved in emotion) was activated only by pleasant aspects. This area has also been shown to represent painful as well as pleasant touch, demonstrating the relevance of this brain region for representing the emotional dimensions of skin sensitivity – both rewarding and punishing (Rolls et al. 2003). Using PET, again with neurologically intact subjects, and our RTS, we stroked the palmar skin and the hairy skin of the forearm and again found that stroking hairy skin activated reward systems in the brain such as the OFC and insular cortex that were not activated when stroking the palm (unpublished data).

These psychophysical, anatomical, behavioural and neuroimaging studies are beginning to help us define a neurobiological and psychological basis for two forms of touch, and to enable us to better interpret the roles of touch in human affiliative and exploratory tactile behaviours.

Exploratory Touch

In somatosensory neurophysiology the canonical view is that touch is mediated by large-diameter, fast-conducting peripheral nerves, and that the sensory acuity of touch across the body is highly heterogeneous, with areas such as the digit tips and the lips being more densely innervated and more cortically represented than other body sites; this explains why the hands and the lips are prominent in tactile exploratory behaviour, such as those performed when manipulating objects or artefacts. With the dominance of the glabrous surface of the hand in all forms of exploratory tactile behaviour and object manipulation (in young infants the lips are often used during exploratory tactile behaviour), it is not surprising that much is known about the function of hand–brain neural systems. However, it is becoming recognized that there is another purpose to touch that is more interoceptive than exteroceptive and which is less accessible to conscious self-report, as described above where CT-afferents have been shown to project to emotional systems but less or not at all towards the discriminative cognitive systems (classical somatosensory areas SI and SII). The 'fast' touch system, comprising the classical LTMs, provides us with a real-time perception of the tactile properties of objects; its high processing speed is essential if we are to handle objects efficiently without dropping or crushing them. The sensory input is one arm of a complex sensorimotor control system that provides millisecond by millisecond updates to the brain to enable the motor systems controlling the hand to work efficiently. The 'slow' touch system, on the other hand, does not feed information in real-time to the conscious brain but provides off-line sub-conscious information to the emotional brain if the object being manipulated is explored by either touching against a hairy skin site – such as the cheek – or via some top-down process if the object is expected to have an emotional salience. To what extent the 'fast' tactile Aß-afferents project to the emotional systems along with the CT-afferents has not been widely explored, but it seems likely that this may be the case because pleasant touch stimuli on the palm, where CT-afferents are lacking, has been shown to activate the orbitofrontal cortex, an part of the brain involved in more complex emotional mechanisms (Francis et al. 1999), and interestingly is itself a target area for nerve fibre projections from the insular cortex. It remains to be established whether a projection of tactile Aß-afferents to emotional centres is unique for the palm or is true for other skin areas as well. It is possible, however, that Aß-afferents from glabrous skin obtain an affective role because CT-afferents are lacking here, or that the widely recognized pleasure experienced by manipulating silk or velvet, for example, represents an aesthetic – learned – response to these materials, based on experiences gained while such surfaces were contacting hairy skin sites. It is clear that affective aspects of touch are not always dependent on CT-afferents alone, but one explanation as to why CTs are not found in glabrous skin is that the hands serve a predominantly discriminative, exploratory and manipulative role that requires 'instant messaging' to be effective, and that would be compromised if 'slow' CT-afferent emotional information is also being processed.

An 'affective touch hypothesis' implies that the essential role of the CT-system is to provide or support emotional, hormonal and behavioural responses to skin-to-skin contact with conspecifics, and to extract and process the pleasant aspects of all types of touch – be they animate or inanimate. It seems likely that a natural perceptive-emotional response to pleasant touch is dependent on the combination of afferents from the two tactile systems, because selective CT-stimulation fails to evoke anything like a full sensation of pleasant touch in patients who have lost their discriminative touch nerves. The combination of bottom-up inputs from CT and Aß-afferents is required for the complete feeling of pleasant touch in hairy skin and the intensity, and even the quality of the emotional response evoked by a particular tactile stimulus, is highly dependent on top-down influences such as context and object history, such as is often the case when there are attributes associated with heritage.

Conclusion

A dual role for touch serving both a discriminative and an affective role in human behaviour has been described. The human hand has clearly evolved to perform a wide range of exploratory and manipulative tasks, and far surpasses this function than in any other primate. Many questions, however, still need to be addressed and answered with regard to the role of the affective CT system in touch behaviour, such as their absence in glabrous skin, which clearly does not preclude us perceiving the pleasant feel of velvet or silk for example. The patients we have studied who only have intact C-fibre systems and no appreciation of cognitive touch are not living lives of pure tactile sensuality; in fact their appreciation of the pleasantness of touch is often diminished compared with responses from 'intact' individuals. The relative importance of hairy skin as opposed to glabrous skin sites for encoding the more emotionally salient aspects of touch may be found by looking at this difference in evolutionary terms where the glabrous skin surfaces, being in direct contact with the ground prior to the emergence of bipedalism in apes, served important thermoregulatory and mechanosensory functions where heat loss from the body was controlled and information for the control of postural muscles during running and walking sensed. In bipedal apes, and eventually humans, the importance of the hands has been critical, particularly in terms of tool use, and it is reasonable to assume that the absence of CTs from glabrous sites serves an adaptive function whereby object handling and manipulation is not confounded by affective information from the palms nor mechanosensory information from the soles of the feet that is used to control gait. Behavioural and neuroimaging studies (PET and fMRI) are addressing these issues and there is increasing evidence for a different central neural representation to stroking either glabrous or hairy skin that can help explain behavioural aspects of touch and touching. With the focus of this chapter on the role of touching and handling objects in a heritage context, these findings should help curators and the general public discover different ways in which to extract 'feeling' from 'touching' artefacts.

It was mentioned in the Introduction that by employing different parts of the body, such as the backs of the hands or the face, or by employing a range of different exploratory tactile movements to achieve different sensory effects, we might be able to enhance the experiences gleaned from interacting with heritage-type objects. Insights into how museums can approach the issue of handling objects in a heritage context – when most of them have serious problems with perceptual access to their exhibits other than with sight – can be provided by studying the recent experience of museums in response to the passing in 1995 of the Disability Discrimination Act, where tactile access to exhibits for blind people was first explored. The first permanent touch tours in a British museum were for the Egyptian and Graeco-Roman sculpture galleries in the British Museum, introduced in 1990, where visually impaired people were allowed to touch particular pieces. Hetherington (2003) describes how a blind colleague went around the gallery touching the sculptures and describing her experiences: 'In part, she was engaged in an act of identification, establishing for herself what each of the sculptures represented. But she was also fascinated by what she called "the feel" and how it connected with her. The textures of the different kinds of stone that had been used and the variations in temperature were all equally important.' Having an acutely developed sense of touch she was able to describe her enhanced sensitivity to the texture of materials in varying ways depending on her tactile strategies such as which hand she used, whether she made broad sweeping movements with her hands, or more local, detailed investigations with her fingertips. Of interest to the aspects of emotional touch described in this chapter, she could also relate experiences of mood and feelings that are best described as 'embodied perception'. Haptic exploration of these objects was much more than the purely discriminative analysis of a museum artefact, it provided proximal knowledge, a knowledge that is difficult to relate verbally as it depends upon implicit non-representational forms of understanding. Hetherington (2003) describes this as a performative approach to issues of understanding the meaning of objects and a 'blurring of the boundaries between the body and the object being handled'. Touch is an embodied sense and as such a source of 'proximal knowledge' (Jospovici 1996; Turner 1992) that sighted people are often not fully aware of, other than instances of affiliative touch where we are aware – in an emotional and often non-verbalizable manner – of its more covert and intimate properties. Somatosensation transcends vision as a source of recognition and understanding and does so without necessarily trying to achieve some representational outcome. Hetherington finds it unsurprising that museums overlook the importance of touch as their heritage is one of visual dominance (Lyotard 1996), and states that proximal knowledge – the intimate handling of objects - has always been the sole privilege of the curator in the museum context. With our increasing knowledge of the neural processes underpinning the more emotional properties of touch, it is to be hoped that museum curators will be better able to find ways to enhance the experiences their visitors derive from museum artefacts.

References

Anderson, D.J. (1999), 'Lineages and transcription factors in the specification of vertebrate primary sensory neurons', *Current Opinion in Neurobiology* 9(5): 517–23.

Bessou, M. and Perl, E.R. (1969), 'Response of cutaneous sensory units with unmyelinated fibres to noxious stimuli', *Journal of Neurophysiology* 32: 1025–43.

Bolanowski, S.J., Gescheider, G.A., Verrillo, R.T. and Checkosky, C.M. (1988), 'Four channels mediate the mechanical aspects of touch', *Journal of the Acoustical Society of America* 84: 1680–93.

Coghill, R.C., Talbot, J.D., Evans, A.C., Meyer, E., Gjedde, A., Bushnell, M.C. and Duncan, G.H. (1994), 'Distributed processing of pain and vibration by the human brain', *Journal of Neuroscience* 14: 4095–108.

Cole, J. (1995), *Pride and a Daily Marathon*, Cambridge, MA: MIT Press.

Craig, A.D. (2002), 'How do you feel? Interoception: The sense of the physiological condition of the body', *Nature Reviews Neuroscience* 3: 655–66.

Craig, A.D. (2003) 'Interoception: The sense of the physiological condition of the body', *Current Opinions in Neurobiology* 13(4): 500–5.

Craig, A.D. and Bushnell, M.C. (2004), 'The thermal grill illusion: Unmasking the burn of cold pain', *Science* 265(5169): 252–5.

Darian-Smith, I. (1984), 'The sense of touch: Performance and peripheral neural processes', in J.M. Brookhart and V.B. Mountcastle (eds) *Handbook of Physiology*, Bethesda, MD: American Physiological Society.

Edin, B.B. (2001), 'Cutaneous afferents provide information about knee joint movements in humans', *Journal of Physiology* 531(1): 289–97.

Essick, G.K. and Edin, B.B. (1995), 'Receptor encoding of moving tactile stimuli in humans. II The mean response of individual low-threshold mechanoreceptors to motion across the receptive field', *Journal of Neuroscience* 15: 848–63.

Essick, G., James, A. and McGlone, F.P. (1999), 'Psychophysical assessment of the affective components of non-painful touch', *Neuroreport* 10: 2083–7.

Francis, S.T., Rolls, E.T., Bowtell, R., McGlone, F., O'Doherty, J.O., Browning, A., Clare, S. and Smith, E. (1999), 'The representation of pleasant touch in the brain and its relationship with taste and olfactory areas', *Neuroreport* 10: 453–9.

Francis, S.T., Kelly, E.F., Bowtell, R., Dunseath, W.J., Folger, S.E. and McGlone, F.P. (2000), 'FMRI of the responses to vibratory stimulation of digit tips', *Neuroimage* 11: 188–202.

Gescheider, G.A., O'Malley, M.J. and Verrillo, R.T. (1983), 'Vibrotactile forward masking: Evidence for channel independence', *Journal of the Acoustical Society of America* 74: 474–85.

Gescheider, G.A., Sklar, B.F., Van Doren, C.L. and Verrillo, R.T. (1985), 'Vibrotactile forward masking: Psychophysical evidence for a triplex theory of cutaneous mechanoreception', *Sensory Processes* 1: 187–203.

Gescheider, G.A., Verrillo, R.T. and Van Doren, C.L. (1982), 'Prediction of vibrotactile masking functions', *Journal of the Acoustical Society of America* 2: 1421–6.

Gescheider, G.A., Bolanowski, S.J. and Verrillo, R.T. (1992), 'Sensory, cognitive and response factors in the judgement of sensory magnitude', in: D Algom (ed.), *Psychophysical Approaches to Cognition*, Amsterdam: Elsevier, pp. 575–621.

Greenspan, J.D. and Lamotte, R.H. (1993), 'Cutaneous mechanoreceptors of the hand: Experimental studies and their implications for clinical testing of tactile sensation', *Journal of Hand Therapy* 6: 75–82.

Hagbarth, K.E. and Vallbo, A.B. (1967), 'Mechanoreceptor activity recorded percutaneously with semi-microelectrodes in human peripheral nerves', *Acta Physiologica Scandinavica* 69(1):121–2.

Harrington, T. and Merzenich, M. (1970), 'Neural coding in the sense of touch: Human sensations of skin indentation compared with responses of slowly adapting mechanoreceptive afferents innervating the hairy skin of monkeys', *Experimental Brain Research* 10: 251–63.

Hertenstein, M., Keltner, D., App, B., Bulleit, B. and Jaskolka, A. (2006), 'Touch communicates distinct emotions', *Emotion* 6(3): 528–83.

Hetherington, K. (2000), Museums and the visually impaired: The spatial politics of access', *Sociological Review* 48(3): 444–63.

Hetherington, K. (2002), 'The unsightly: Visual impairment, touch and the Parthenon Frieze', *Theory, Culture and Society* 19(5/6): 187–205.

Hetherington, K (2003) 'Accountability and disposal: Visual impairment and the museum', *Museum and Society* 1(2): 104–15.

Iggo, A. and Korhuber, H.H.A. (1977), 'A quantitative study of C-mechanoreceptors in the hairy skin of the cat', *Journal of Physiology* 271: 549–65.

Jarvilehto, T., Hamalainen, H. and Laurinen, P. (1976), 'Characteristics of single mechanoreceptive fibres innervating hairy skin of the human hand', *Experimental Brain Research* 25: 45–61.

Johansson, R.S. (1976), 'Receptive field sensitivity profile of mechanosensitive units innervating the glabrous skin of the human hand', *Brain Research* 219: 13–27.

Johansson, R.S., Trulsson, M., Olsson, K.A. and Westberg, K.G. (1988), 'Mechanoreceptor activity from the human face and oral mucosa', *Experimental Brain Research* 72: 204–8.

Jospovici, G. (1996), *Touch,* New Haven, CT: Yale University Press.

Katz, D. (1989[1925]), *The World of Touch*, Hillsdale, NJ: Lawrence Erlbaum.

Liu, Q., Vrontou, S., Rice, F., Zylka, M.J., Dong, X. and Anderson, D.J. (2007), 'Molecular genetic visualisation of a rare subset of unmyelinated sensory neurons that may detect gentle touch', *Nature Neuroscience* 10(8): 946–8.

Loken, L.S., Wessberg, J., McGlone, F., Olausson, H.W. (2006), 'The response of human unmyelinated tactile afferents to brush stroking with varying velocity', *Society for Neuroscience* Abstracts (poster 143.5).

Lyotard, J-F. (1996), 'Les immateriaux', in R. Greenberg, B. Ferguson and S. Nairne (eds), *Thinking About Exhibitions*, London: Routledge, pp. 159–73.

McGlone, F.P., Kelly, E.F., Trulsson, M., Francis, S.T., Westling, G. and Bowtel, R. (2002), 'Functional neuroimaging studies of human somatosensory cortex', *Behavioural Brain Research* 135: 147–58.

McGlone, F., Vallbo, A., Olausson, H., Loken, L. and Wessburg, J. (2007), 'Discriminative and emotional touch', *Canadian Journal of Experimental Psychology* 61(3): 173–83.

Maeda, K., Kakigi, R., Hoshiyama, M. and Koyama, S. (1999), 'Topography of the secondary somatosensory cortex in humans: A magentoencephalographic study', *Neuroreport* 10: 301–6.

Maldjian, J.A., Gotschalk, A., Patel, R.S., Detre, J.A. and Alsop, D.C. (1999), 'The sensory somatotopic map of the human hand demonstrated at 4T', *Neuroimage* 10: 55–62.

Mesulam, M. (2000), '*Principles of Behavioural and Cognitive Neurology*, 2nd edn, Oxford: Oxford University Press.

Mountcastle, V.B. (1984), 'Neural mechanisms in somesthesia', in V. B. Mountcastle (ed.), *Medical Physiology,* St Louis, MO: Mosby, pp. 307–47.

Nordin, M. (1990), 'Low threshold mechanoreceptive and nociceptive units with unmyelinated (C) fibres in the human supraorbital nerve', *Journal of Physiology* 426: 229–40.

Olausson, H., Lamarre, Y., Backlund, H., Morin, C., Wallin, B.G., Starck, S., Strigo, K., Worsley, K., Vallbo, A.B. and Bushnell, M.C. (2002), 'Unmyelinated tactile afferents signal touch and project to the insular cortex', *Nature Neuroscience* 5: 900–3.

Pecina, S., Smith, K. S. and Berridge, K. (2006), 'Hedonic hot spots in the brain', *Neuroscientist* 12(6): 500–11.

Penfield, R. and Boldrey, E. (1937), 'Somatic motor and sensory representation in the cerebral cortex of man as studied by electrical stimulation', *Brain* 60: 389–443.

Penfield, R. and Rasmussen, T. (1952), *The Cerebral Cortex of Man*, New York: Macmillan.

Reilly, D.M., Ferdinando, D., Johnston, C., Shaw, C., Buchanan, K.D. and Green, M. (1997), 'The epidermal nerve fibre network: Characterization of nerve fibres in human skin by confocal microscopy and assessment of racial variations', *British Journal of Dermatology* 137: 163–70.

Rolls, E., O'Doherty, J., Kringelbach, M., Francis, S., Bowtell, R. and McGlone, F. (2003), 'Representation of pleasant and painful touch in the human orbitofrontal cortex', *Cerebral Cortex* 10(3): 284–93.

Schneider, R.J., Friedman, D.P. and Mishkin, M. (1993), 'A modality specific somatosensory area within the insula of the rhesus monkey', *Brain Research* 621: 116–20.

Serra, J., Campero, M., Ochoa, J. L. and Bostock, H. (1999), 'Activity-dependent slowing of conduction differentiates functional subtypes of C fibres innervating human skin', *Journal of Physiology* 515: 799–811.

Trulsson, M., Francis, S.T., Kelly, E.F., Westling, G., Bowtell, R. and McGlone, F.P. (2000), 'Cortical responses to single mechanoreceptive afferent microstimulation revealed with fMRI', *Neuroimage* 11: 188–202.

Turner, B. (1992), 'Reflections on the epistemology of the hand', in B. S. Tuner (ed.), *Regulating Bodies*, London: Routledge, pp. 99–121.

Vallbo, A.B., Hagbarth, K., Torebjork, E. and Wallin, B.G. (1979), 'Somatosensory, proprioceptive snd sympathetic activity in human peripheral nerves', *Physiological Reviews* 59: 919–57.

Vallbo, A.B., Olausson, H., Wessberg, J. and Norsell, U. (1993), 'A system of unmyelinated afferents for innocuous mechanoreception in the human skin', *Brain Research* 628: 301–3.

Wessberg, J., Olausson, H., Fernstormm, K.W. and Vallbo, A.B. (2003), 'Receptive field properties of unmyelinated tactile afferents in the human skin', *Journal of Neurophysiology* 89: 1567–75.

Westling, G.K. (1986), *Sensori-motor Mechanisms During Precision Grip in Man,* Umeå University Medical Dissertations, New Series 171, Umeå Sweden: Umeå University.

Willis, W.D. (1985), *The Pain System*, Basel: Karger.

Willis, W.D. and Coggeshall, R.E. (1991), *Sensory Mechanisms of the Spinal Cord,* second edn, New York: Plenum Press.

Woolsey, C. (1946), 'Second somatic receiving areas in the cerebral cortex of the cat, dog and monkey', *Fed. Proc.* 2: 55–6.

Zottermann, Y. (1939), 'Touch, pain and tickling: An electrophysiological investigation on cutaneous sensory nerves', *Journal of Physiology* 95: 1–28.

–4–

Emotional Touch: A Neuroscientific Overview

Hugo Critchley

Skin is a main interface between an individual and the outside world. The sense of touch informs and guides close interactions with the immediate environment. The skin is also the largest organ in the body and is vital for controlling and maintaining the health of the body as a whole, a process known homoeostasis, to ensure the stability of the internal environment. Emotions arise from the demands of meeting inner bodily needs – for example, in hunger or thirst – and the skin is not left out of these emotional processes. Touch, as an exploratory sense, is complemented by parallel sensations intrinsic to the skin that have strong motivational meaning. Particular nerve fibres carry these allied sensations (including sensual touch, tickle, temperature and pain) to the brain. The fibres themselves are of the same type as the nerves that innervate and control the internal bodily organs. These visceral and skin sensations reach the brain by a different pathway to conventional, discriminatory, analytic touch. Within the brain, the information by-passes somatosensory cortex to access directly emotional regions that support emotional feeling states, guide and motivate behaviour and reinforce memories.

The Hypnoglyph

There is a short science fiction story that I recall reading at school entitled 'The Hypnoglyph' (Anthony 1953). I recently located an original copy of the anthology, and also learned that the author was John Anthony Ciardi, an American poet and academic, who wrote this science fiction piece in 1953 under the name John Anthony. The story involves a man visiting a friend who collects unusual alien artefacts, including a beautifully carved piece of wood. The owner calls it the prize of his collection and hands it over to his guest to feel and stroke. They chat about tactile pleasure and discuss a late twentieth-century artist and sculptor whose work focused on creating sensual touch experiences. Then the collector begins the story about how he acquired the carved object. He describes prospecting for crystals in deep space when his team discover a new planetary system that previously was thought merely to be an asteroid belt. On one of the planets there is intelligent primate-based life. There he retrieves the object that he says, surprisingly, was a hunting implement. He describes the human-like alien culture further:

> The women of the proper clan work these things and the men set them out in the forests ... the animals with their extremely high tactile suggestibility come through the forest and find one of these things in their way. They begin to stroke and feel it and they just can't stop. The men don't even kill them, all the slaughtering is done by the ruling clan of women. The men simply wait until the animal has worked itself into a right state then lead it back to the slaughtering compound. (Anthony 1953)

I think the notion of a hypnoglyph encapsulates many concepts relating to emotional touch and object handling: certain types of stimulation to the skin may convey a positive emotional power that engages, calms and sedates in a way that contrasts with unpleasant emotional sensations such as pain and ache, and that differs categorically from tactile spatial discrimination.

The Emotional Brain and Peripheral Theories of Emotion

My research has focused on emotional processes and has largely steered clear of somatosensory pathways (the long-established neural substrates for sensation, including pain and discriminatory touch) and motor pathways mediating our volitional and involuntary interactions with the world. I have tended to concentrate on areas of the brain involved in motivation, reward and the emotional influence on behaviour. These 'emotional' brain regions are named after their appearance or location: thus the amygdala is so-called because of its almond shape, the cingulate cortex is a belt of tissue along the inner surface of the hemispheres, the orbitofrontal cortex lies above the eyes at the front of the brain and the insula cortex is an 'island' of tissue tucked in between the front and side of the brain (see Figure 4.1).

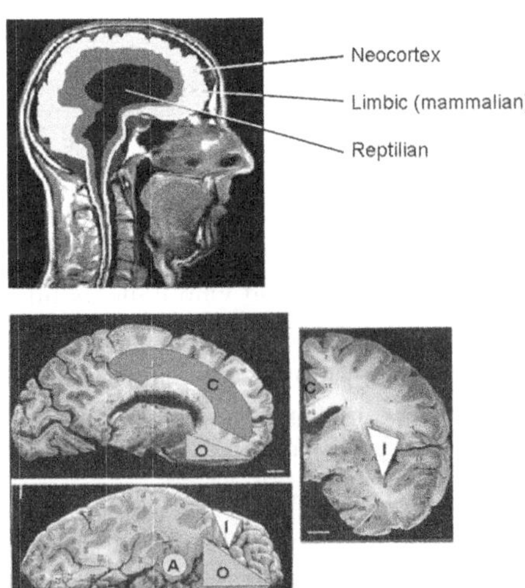

Figure 4.1 Schematics of the emotional brain (McLean 1954).

The upper illustration in Figure 4.1 shows Maclean's concept of the triune brain. A brainstem-based reflexive 'reptilian' brain evolved into a 'limbic system'-based 'mammalian' brain that supports affective behaviour and emotional feeling states. The logical reasoning brain is seated in the overlying neocortex that has evolved exponentially in humans.

Much of the recent focus of human studies into the brain systems supporting emotion has focused on a few areas, which are illustrated in the lower panel – namely, A: Amygdala; C: Cingulate gyrus; I: Insula cortex; O Orbitofrontal cortex. The amygdala and cingulate are historically components of the limbic system, while orbitofrontal cortex and insula are viewed as 'paralimbic', nevertheless old, cortical regions.

The function of these brain regions is closely linked to the inner workings of the visceral organs whose control is maintained at the lowest reaches of the brain (the brainstem). They are evolutionarily old, present in the brains of all mammals, but pre-dating the massive growth of overlying 'neocortex' that has occurred in the brains of primates, particularly humans (Dalgleish 2004). The contribution of these brain regions to mood, motivation and memory has been conceptualized in terms of a system or circuit – for example, as the 'Papez circuit' (after anatomist James Papez, (pronounced *Paips*) and as the 'limbic system', since these structures form a rim (Latin *limbus*) around the older parts of the brain). The term 'mammalian brain' has also been used as an alternative (Papez 1937; McLean 1954). Essentially, this emotional brain is distinguishable from 'higher level' neocortical regions that are necessary for logical thought and deductive reasoning and declarative perception (being able to demonstrate awareness). The emotional brain is also a level up from the brainstem and hypothalamic centres involved in reflex responses that characterize and encapsulate the behaviour of reptiles (species not known for their emotional warmth), but is nevertheless influenced by the motivations arising from the day-to-day unconscious 'homoeostatic' running of the body.

Psychological sciences have tended to subscribe to a Cartesian duality, separating mind from body, or perhaps more recently brain from body. However, this stance ignores a wealth of evidence indicating the interdependence of brain and bodily processes. At its most basic, and perhaps obviously, we treat things in our environment differently according to the motivational priorities set by our body. Food is perceived, eaten or rejected and remembered differently depending on whether or not we are hungry – the *feeling state* arising from our body's need for nutrients. Similarly, the basic emotions of thirst, cold and well-being also reflect the brain's representation of our internal bodily state. When events or stimuli engender changes in our bodily state, they have emotional consequences – if we see something that we love or fear, our hearts skip a beat or our palms sweat. If something we encounter (or think) does not automatically elicit a bodily reaction along the lines of an arousal response or shift to different 'vegetative' state, then it is difficult to argue that the stimulus has any emotional potency. This was the logic of a previously unfashionable reductionist theory of emotion, proposed in the 1880s by the American psychologist William James and the Danish physician/psychologist Carl Lange. The James-Lange theory of emotion states that affective experience – that is, emotional feelings and the 'emotional colour' of events – derive from the re-representation within the brain of changes in the body (James 1894; Lange 1885) (see Figure 4.2). Lange emphasized the contribution to emotion of changes in the viscera, blood vessels and internal organs. These are diffuse sensations that are difficult to describe or localize. The theory has various implications, including the notion that different bodily reactions accompany (and predict) different emotional feelings.

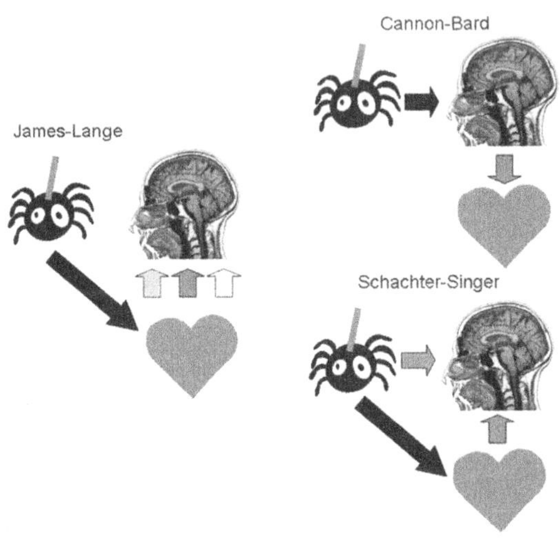

Figure 4.2 Concepts of emotion.

The left side of Figure 4.2 depicts the James-Lange theory of emotion, which proposes that what is different about emotions is change in the arousal state of the body. When we see an object, emotional colour (feeling) is provided by feedback from the body, and this is illustrated diagrammatically – the mind's perception of the heart rate change, rather than the spider, determines if an emotion is experienced. Different emotions ought to be associated with different body states.

In 1929 Cannon criticized the peripheral theory of James and Lange on a number of counts, not least that bodily arousal was too undifferentiated to account for the wealth of different emotions. The Cannon-Bard theory, as it is known, is depicted in the upper right-hand part of Figure 4.2. They argued that the processing of emotion was solely within the mind and the bodily response if present was an epiphenomenological result of this processing. In 1962 Schachter and Singer (1962) argued that 'automatic' bodily arousal could provide the intensity of emotional experience, but its co-occurrence with a cognitive perception determined which emotion was experienced. This view, which is widely held today, is depicted in the lower right part of Figure 4.2.

The James-Lange theory has been hotly debated over the years, not least by Walter Cannon who, in the 1920s, argued among other things that bodily responses were too undifferentiated to be emotion-specific (Cannon 1929) (see top right Figure 4.2). However, the idea that peripheral autonomic arousal contributes to emotional intensity is relatively robust, as illustrated in the 'two-stage' model of emotion proposed by Schachter and Singer (1962) which argues that the cognitive context in which bodily arousal is perceived determines the quality of emotion: Schachter and Singer injected participants with adrenaline and put them in a room with either an irritable or happy actor, the participants' subsequent arousal-induced mood reflecting that of the actor (see bottom right, Figure 4.2). Thus, the adrenaline induced arousal means to the participant, 'I have an emotion', the context (happy actor) means the arousal is interpreted is 'I have an emotion and it is happiness'. This arousal-plus-interpretation account of emotion is widely held by psychologists.

In recent years, there has been a resurgence of interest in emotional processes, including this dynamic link between brain and body, and much credit is due to the research and writings of Antonio Damasio (1999). Damasio and others have proposed a hierarchical structure to emotional states which ultimately govern survival through responding to, or anticipating, the homoeostatic, defensive or reproductive needs of our body. There has also been some progress in showing that different patterns of autonomic bodily arousal accompany different emotions, consistent with predictions from the James-Lange theory (Levenson et al. 1990; Critchley et al. 2005; Rainville et al. 2006).

The autonomic nervous system provides a major route for the brain to control the internal state of the body. It has two main divisions, the sympathetic and parasympathetic systems, which have different functions. Heuristically, the sympathetic division drives fight-or-flight responses that speed heart rate, increases blood pressure and diverts blood to muscles. Conversely the parasympathetic autonomic division represents a 'rest-and-digest' (vegetative) system, slowing heart rate and diverting blood to the gut. The balance between sympathetic and parasympathetic systems sets the tone of arousal in the body. The information regarding the internal state of the body travels back to the brain along the same nerve sympathetic and parasympathetic bundles via the spinal cord and vagus nerve. A.D. (Bud) Craig, a scientist based in Arizona, has shown that in the spinal cord and the brain, the pathways taken by these feedback signals from the bodily organs are very specialized and different from the routes taken by sensory signals such as discriminatory touch and joint position and certain aspects of pain (Craig 2003).

Imaging Feelings

In recent years, brain imaging research has helped neuroscientists and psychologists to identify when, where and how in the human brain thoughts and emotions occur. These brain-scanning methods have complemented other types of neuroanatomical investigation to tease apart the brain systems involved in emotion. Based on the simple peripheral model of emotion, we have sought to identify which brain systems are involved in generating changes in autonomic bodily arousal that accompany different behaviours and, correspondingly, which brain regions seem to represent and map changes in the internal autonomic bodily arousal state occurring during emotional behaviour. This work has relied on correlating activity in different parts of the brain with measurements of bodily responses such as heart rate, blood pressure, skin conductance (reflecting sweat gland activity) and pupil dilatation. In addition, we have used patients with brain damage or problems with the autonomic nervous system to validate our hypotheses and provide a more detailed account of our observations. The data from these studies argue that a set of brain regions (namely the anterior cingulate cortex, parts of the frontal lobe and, in the case of emotional processing, the amygdala) are responsible for driving autonomic arousal during effort and emotion (Critchley 2005).

Different regions of frontal lobe (orbitofrontal cortex) and ventral anterior cingulate are involved in suppressing sympathetic arousal, and perhaps driving parasympathetic responses to different emotional behaviours (Nagai 2004). In a study of patients with autonomic failure (who cannot generate a change in autonomic bodily state) changes were seen in the responses of the amygdala and insula cortex, suggesting that these areas also represent internal bodily arousal (Critchley et al. 2002). The responses within a region called insula cortex seemed to be particularly involved in combining information about the arousal state of the body with emotional information that can be consciously perceived (e.g. threat). This kind of integrative processing is in keeping with what Schachter and Singer (1962) predicted the brain was doing to give rise to generate emotional feelings.

The same insula cortex region was also implicated in a study of 'interoceptive awareness'. This study tested the prediction that people who are more aware of their own internal bodily arousal states would experience emotions with greater intensity. A 'heartbeat detection task' is the kind of task used by psychologists to determine if people are more or less aware of their own bodily changes. For example, the participant hears a sequence of ten beeps played through a set of headphones. At the end of each set of ten notes (these are repeated) they have to say if the notes were in time with their own heartbeats, or came after their own heartbeats (a short delay was added on some occasions). We scanned people when they performed such a heartbeat detection task. The results showed that activity within the insula cortex reflected people's conscious awareness of internal bodily responses (i.e. how accurately they judged the timing of their own heartbeats). We also found that this activity predicted the participants' day-to-day experiences of emotions such as anxiety. This study fits with other studies that suggest people who are naturally more aware of their bodily responses, including heartbeat, experience heightened emotions (in some cases for the worst, if the main emotion is anxiety). It also suggests that one particular brain region, the insula cortex, could be the 'seat' of emotional feelings arising from internal bodily sensations (Critchley et al. 2004).

When people try to judge the timing of their heart beat they activate a number of brain areas, notably the insula cortex. Figure 4.3 (upper left) shows the average activity of a group of people performing a heart beat detection task (Critchley et al. 2004).

Conscious awareness of internal arousal (i.e. how well people know the timing of their own heart beat) is predicted by the activity within the insula, and the lower left panel shows a brain picture of group activity. The more active the insula and/or the more sensitive one is to heart rate, then the more one experiences day-to-day anxiety and other emotional states, and this is illustrated in the graph.

Olausson et al. (2002) showed that stimulation of emotional touch fibres activates brain areas, particularly the insula cortex. This was even demonstrated in patients with no neurological condition of myelinated touch fibres. By stroking their arms, specific emotional touch fibres only could be activated, producing conscious sensory feelings associated with brain activity in the insula. The brain picture in the right-hand panel of Figure 4.3 is from such a patient who had no myelinated touch fibres and therefore there is no activation of somatosensory cortex, but insula activity when emotional touch nerves are activated when stroked.

Emotions ultimately drive our behaviour. They are linked to reward and punishment and are directed by feedback from the body regarding our motivational and survival needs. Emotions are a system for 'prioritizing evolutionary salience'. If our evolutionary ancestors had not felt fear they would not have run away and we would not be here. The brain systems that give rise to emotional feelings are enmeshed within the control systems that control the health and functioning of the body (and which give rise to motivations, as mentioned above). Emotional feeling states arise from conscious awareness of these processes through their representation within specific brain regions, such as the insula cortex.

Skin

The skin is often declared, almost paradoxically, to be the largest organ in the body. The skin is the primary interface of an individual with the nearby environment. It is a sensory organ through which the environment is explored directly by touch. The skin is also essential for homoeostasis, the main role of the internal 'visceral' organs. Skin controls temperature, fluid balance, energy storage and vitamin synthesis. Aspects of touch, including sensual touch, tickle and pain appear more related to visceral afferent feedback (from internal bodily organs and bodily arousal systems) than classical somatosensory (discriminatory touch) sensation. Because of this, touch can be motivationally salient, rewarding – that is, what learning theorists term a primary reinforcer (something that encourages certain behaviours): for instance, infants calm down when held.

The work of Elliott Blass and others has indicated that the pain threshold of infants may be altered by holding the child (Gray et al. 2000). Holding or giving milk or a sweet taste will also calm an infant and lower heart rate (Blass and Watt 1999). The effect of sweet tastes is long lasting (and in fact blocked by opiate antagonists or attenuated in children of heroin addicts) suggesting that it depends on endogenous endorphins. The effect of touch and holding appears to use a different chemical system.

In other primates (perhaps a bit more than in humans), grooming other individuals is an essential feature of socialization and social behaviour. In 1996, Kosei Taira (Taira and Rolls 1996) published the results of an interesting experiment in which he had spent almost two years proving that, for monkeys, grooming is rewarding. In his experiment, a monkey heard a tone and then had to touch a picture, either a square or a cross. If the monkey touched the correct picture, it received a reward. In this case the reward was being briefly groomed around the head and neck by Dr Taira. The monkey learned how to do this task, selecting the correct picture in order to be stroked around the head and neck. When the task was 'reversed' (i.e. when the learned association between touching the square and receiving grooming was changed to touching the cross to receive grooming), the monkey became quite agitated for a few days, indicating that it found this type of grooming rewarding. The reversal change of the correct picture was eventually relearned with practice by the monkey. Overall, the experiment highlighted that a particular type of touch (grooming) directly activates the reward systems of the brain to control behaviour. Thus emotional touch has strong motivational incentive values.

Emotional Touch Pathways Identified in Humans

I think we all appreciate that pain can be a type of 'touch' sensation with profound emotional meaning. Rubbing the skin (or acupuncture stimulation) may block pain signals through what are now well-established mechanisms (Melzack and Wall 1965). Pain sensations are in part conveyed by small *unmyelinated* nerve fibres (nerve fibres lacking a myelin sheath) called C-fibres. Myelin around nerves provides 'electrical' insulation to speed up signalling, thus information travels slowly along these unmyelinated C-fibres. There are some faster *myelinated* pain fibres in the skin (known as Aδ-fibres) mediating rapid reflex withdrawal in response – for example, if I burn my hand. The sensation conveyed by these Aδ-fibres is typically sharp. In contrast, C-fibre stimulation tends to produce a deep badly localized ache. In 1993

Åke Vallbo and colleagues reported a different sort of sensory nerve in the skin: an unmyelinated 'C-fibre' nerve conveying non-pain signals from the skin. Their paper comments "Their biological role remains an enigma which might attract more attention now that their existence in man has been demonstrated" (Vallbo et al. 1993).

The work of a number of individuals, described in more detail in the chapter by Francis McGlone (this volume), has highlighted the role of these nerve fibres in conveying emotional touch to the brain. Emotional touch describes tactile sensations, including sensual stroking, pressure, vibration, itch and tickle, that can be pleasant or unpleasant. Brain-imaging studies have recently shown that the brain centres activated by pleasant and unpleasant 'emotional touch' are not the same as other forms of touch. In fact they are the same brain regions involved in emotional representations and visceral feeling states (including the insula cortex) and not the regions of the somatosensory cortex on the lateral surface of the brain that support tactile discrimination.

A landmark study was undertaken by H. Olausson and co-workers. They examined patients who had totally lost their conventional sensation of touch (Olausson et al. 2002). Smooth, soft, slow stimulation of the forearms of these patients elicited pleasant stroking sensations through the activation of the unmyelinated, non-pain nerve fibres, in some cases to the great surprise of the patient, illustrating the integrity of their emotional touch. During brain imaging of these patients, emotional centres were activated, notably including the insula cortex, consistent with the direct transmission of these skin sensations into centres that govern how we feel emotionally.

Conclusion and 'The Hypnoglyph'

This is a brief account of emotional touch and its relationship to emotional systems. Emotional feeling states are integrated with the control and feedback regulation of the internal bodily state. This relationship couples internal signals to motivation and survival related behaviours. There is a type of touch which taps directly into this system, with both negative (painful) and positive components. This emotional touch appears to be important for infant rearing, social interaction, pair bonding and, presumably, sex in humans. Sensual touch has calming effects on autonomic and behavioural arousal enhancing physiological as well as subjective well-being. Importantly, the exploration of our environment through touch is coupled to motivational and emotional systems within the brain. The process of object handling to derive a sense and feel for the item can imbue the object with character, significance and emotional colour in a manner perhaps independent of functionality. The broader comprehension gained through object handling may ultimately be more satisfying, especially if the object is intrinsically pleasant to feel and hold.

To return to the story 'The Hypnoglyph', the man visiting his friend is sitting by the table, holding a carved object picked from the table. He is told that the object is used to hunt animals for food and that it works by hypnotizing them through the power of touch. The visitor becomes more relaxed and drowsy as he listens to the story. The last words he hears are his host calling to his alien wife "*He is ready darling*".

References

Anthony, J. (1953), 'The Hypnoglyph', *Fantasy and Science Fiction* 5: 25–32.

Blass, E.M. and Watt, L.B. (1999), 'Suckling- and sucrose-induced analgesia in human newborns', *Pain* 83: 611–23.

Cannon, W.B. (1929), *Bodily Changes in Pain, Hunger, Fear and Rage: An Account of Recent Research into the Function of Emotional Excitement*, 2nd edn, New York: Appleton.

Craig, A.D. (2003), 'Interoception: The sense of the physiological condition of the body', *Current Opinion in Neurobiology* 13: 500–5.

Critchley, H.D. (2005), 'Neural mechanisms of autonomic, affective, and cognitive integration', *Journal of Comparative Neurology* 493: 154–66.

Critchley, H.D., Mathias, C.J. and Dolan, R.J. (2002), 'Fear conditioning in humans: The influence of awareness and autonomic arousal on functional neuroanatomy, *Neuron* 33: 653–63.

Critchley, H.D., Rotshtein, P., Nagai, Y., O'Doherty, J., Mathias, C.J. and Dolan, R. (2005), 'Activity in the human brain predicting differential heart rate responses to emotional facial expressions', *Neuroimage* 1: 751–62.

Critchley, H.D., Wiens, S., Rotshtein, P., Ohman, A. and Dolan, R.J. (2004), 'Neural systems supporting interoceptive awareness', *Nature Neuroscience* 7: 189–95.

Dalgleish,T. (2004), 'The emotional brain', *Nature Reviews Neuroscience* 5: 582–9.

Damasio, A.R. (1999). *The Feeling of What Happens: Body and Emotion in the Making of Consciousness*, New York: Harcourt Brace.

Gray, L., Watt, L. and Blass, E.M. (2000), 'Skin-to-skin contact is analgesic in healthy newborns', *Pediatrics* 105: e14.

James, W. (1994[1894]), 'Physical basis of emotion', *Psychological Review* 1: 516–29; reprinted in 1994. *Psychological Review* 101: 205–10.

Lange, C.G. (1885), 'The mechanism of the emotions', in B. Rand (ed.), *The Classical Psychologist,* Boston: Houghton Mifflin, pp.672–85.

Levenson, R.W., Ekman, P. and Friesen, W.V. (1990), 'Voluntary facial action generates emotion-specific autonomic nervous system activity', *Psychophysiology* 27: 363–84

McLean, P.D. (1954), 'The limbic system and its hippocampal formation: Studies in animals and their possible application to man', *Journal of Neurosurgery* 11: 29–44.

Melzack, R. and Wall, P.D. (1965), 'Pain mechanisms: A new theory', *Science* 50: 971–9.

Nagai, Y., Critchley, H.D., Featherstone, E., Trimble, M.R.,and Dolan, R.J. (2004), 'Activity in ventromedial prefrontal cortex covaries with sympathetic skin conductance level: A physiological account of a "default mode" of brain function', *Neuroimage* 21: 1232–41.

Olausson, H., Lamarre, Y., Backlund, H., Morin, C., Wallin, B.G., Starck, G., Ekholm, S., Strigo, I., Worsley, K., Vallbo, A.B. and Bushnell, M.C. (2002), 'Unmyelinated tactile afferents signal touch and project to insular cortex', *Nature Neuroscience* 5: 900–4.

Papez, J.W. (1995 [1937]), 'A proposed mechanism of emotion', *Journal of Neuropsychiatry and Clinical Neuroscience;* reprinted in 1995: 103–12.

Rainville, P., Bechara, A., Naqvi, N. and Damasio, A.R. (2006), 'Basic emotions are associated with distinct patterns of cardiorespiratory activity', *International Journal of Psychophysiology* 61(1): 5–18.

Schachter, S. and Singer, J.E. (1962), 'Cognitive, social, and physiological determinants of emotional state', *Psychological Review* 69: 379–99.

Taira, K. and Rolls, E.T. (1996), 'Receiving grooming as a reinforcer for the monkey', *Physiology and Behavior* 59: 1189–92.

Vallbo, A., Olausson, H., Wessberg, J. and Norrsell, U. (1993), 'A system of unmyelinated afferents for innocuous mechanoreception in the human skin', *Brain Research* 628: 301–4.

Part II
New Technologies for Enhancing Object Interpretation

–5–

The Use of Haptic Interfaces in Haptics Research

Christos Giachritsis

Arguably, touch could be considered as the ultimate sense which allows us to build a complete representation of the world. For example, a sound may attract our attention and make us turn to the general direction of the object or event that caused it. The image of the object or event, projected onto our retina, will establish its exact spatial location and provides us with more details about the parts involved. It may also encourage us to approach and make contact with the object or event. However, only by touching the physical object (or objects) that caused the event, can we establish its physical presence.

Investigating how the sense of touch works is particularly challenging given the nature of the stimulus, which has to be as real as possible. Auditory and vision researchers have access to high-fidelity interface systems that can be used to generate distal stimuli very convincingly, through simulation of physical properties (e.g. head-mounted displays, 3D displays with/out stereo goggles, 3D volumetric displays). Due to this capability of the auditory and visual technologies the proximal stimulus is also very effective. However, haptic interfaces are less advanced and are less capable of generating effective distal stimuli that can explore the entire capacity of the haptic system. Nonetheless, advances in haptic technologies have allowed the generation of certain types of stimulation which can explore part of the haptic system. This technological progress has allowed haptics researchers to create an integrated computerized tool which combines presentation of virtual stimuli, research methods and data collection.

This chapter presents a brief account of how modern haptic interfaces can be used in haptics research to advance our knowledge about how the sense of touch works. The first section introduces fundamental psychophysical concepts and methods which are used in haptics research. The second section presents some of the most popular haptic interfaces which are used in current haptics research and the third section gives an example of a haptics study that combines a haptic interface with psychophysical methods to measure human haptic sensitivity to the specific geometrical property of surface orientation.

Haptics Research

What Haptics Research Does

The term 'haptics' derives from the Greek word *απτικός* (*hapticos*) which means tactile. Therefore, haptics research is the discipline of studying how human beings gain information and knowledge about the tactile world through the haptic system.

What Information Is Collected?

The haptic system combines both *cutaneous* and *kinaesthetic* (from the Greek word *κίνηση* ('kinese', which means 'movement') signals to help us build an effective representation of a physical object (Schiffman 2000). Kinaesthetic signals provide us with information about the movement and position of our limbs while cutaneous signals provide us with information about the nature of the surface of the object. For example, when we reach out to grab an object with our hand, kinaesthetic signals from the wrist, elbow and shoulder joints inform us about the spatial location of the object relative to our body. Once we have made contact with the object and start manipulating it, kinaesthetic signals from our finger joints provide us with cues about its size and shape. Finally, cutaneous signals from skin contact transfer information about material properties and material state, such as surface texture (e.g. rough, smooth), weight (e.g. heavy, light), and temperature (e.g. cold, hot). Thus, the haptic system is equipped with mechanisms that allow us to perceive fundamental physical variables, such as its direction and distance from the body, geometry, texture, weight and temperature. The purpose of gathering all this information is to build a reliable percept of that object.

How Is Information Collected?

Unlike other sensory systems, such as the auditory and visual system, which are gathering information about physical properties and events in a passive way, the haptic system can perceive physical properties and events both passively and actively. Some physical properties, such as texture, are accurately perceived through both passive and active touch (Johnson and Hsiao 1992; Verillo, Bolanowski and McGlone 1999), while other properties, such as weight, are best perceived through actively lifting an object with our hand rather than just resting the object on our palm (Brodie and Ross 1984). Even though passive touch allows us to perceive some object properties, active touch is considered to be the main advantage of the haptic system since it allow us to explore and perceive rather than simply perceive the environment. For example, when we explore an object, the cutaneous and kinaesthetic signals that we receive depend both on our movements and the properties of the object, such as size, shape and texture (Gibson 1962). Nonetheless, as in the case of all sensory modalities, the active touch gathers only environmental information that is useful for the biological system as a whole. Therefore, the limited resources of the haptic system are tuned to explore only the type of information that can be useful to complete important tasks successfully.

Psychophysical Methods

In order to determine what type of physical information is important and understand how the haptic system gathers this information, researchers use a variety of methods and tools. In principle, to investigate how accurately the haptic system perceives object properties such as size, shape, orientation, weight and temperature, scientists use methods of *psychophysics*. Psychophysics is the scientific discipline that attempts to find the rules that govern the relations between external physical properties, or events (the distal stimuli), the stimuli that they generate (the proximal stimuli) and the sensations they cause when interacting with the different sensory systems[1]. The primary aim of psychophysics is to determine *absolute thresholds* (AT) and *difference thresholds* or *difference limen* (DL) of tactile sensitivity. Absolute threshold tells us about the minimum level of a stimulus that has to be sensed in order to perceive the corresponding physical property. For example, the minimum vibrations, transmitted through an object that are detectable to human hand range between 150 and 200 Hz (Brisben et al. 1999). Now, given a detectable stimulus level, difference threshold tells us about the minimum change (increase or decrease) in stimulus intensity that is necessary to produce a *just noticeable difference* (Gescheider 1997). For example, the DL for weight perception at a given weight of 50 g is approximately 6 g (Ross and Brodie 1987). That is, someone can reliably discriminate between 50 g weight and a heavier (but not lighter) than 56 g weight. In all sensory modalities, the relationship between a given stimulus intensity and the magnitude of the difference threshold can be expressed by a constant function, which is know as *Weber's law*. Therefore, given a starting stimulus intensity I_0 and a new intensity I_1, which is just noticeably different, Weber's law states that:

$$(I_0 - I_1)/I_0 = k$$

where k is a constant.

The most common psychophysical paradigms which are used to investigate absolute and difference thresholds of the human haptic system are the *adjustment*, *constant stimuli* and *staircase* methods.

Adjustment

The method of adjustment requires the participant to actively adjust (or reproduce) a physical property. In the case of absolute threshold, the participant is asked to adjust (usually by using a purpose-built apparatus) the intensity level of the stimulus until the stimulus is perceived or disappears. For example, say that an experimenter is interested in finding out what is the absolute threshold of sensitivity to passive perception of a weight that rests on the palm of our hand. He or she may ask the participant to bring a weight in contact with their palm (without visual feedback) until they sense it. The participant is asked to repeat this process for a number of times and then the absolute threshold is taken to be the average value across all trials (Figure 5.1a). In the case of difference threshold, the participant is asked to explore a test tactile stimulus and then adjust a comparison stimulus until it appears to be the same as the test stimulus. For example, say that the researcher investigates sensitivity to tactile length. They may ask the participant to explore a *standard* (STD) length under specified experimental conditions (e.g. explore with one hand in the *x*-axis) and then to adjust (e.g. with one or two hands depending on the experimental condition) a test length or indicate on a ruler (or by other means) the length of the rod. This process is repeated for *N* times to obtain a normal distribution for the reproduced length. The mean of the normal distribution is referred to as *Point of Subjective Equality* (PSE) and represents the value which is subjectively equal to the STD stimulus. The standard deviation of the normal distribution is taken as the difference threshold for that length (Figure 5.1b) (Gescheider 1997; Wetherill and Levitt 1965).

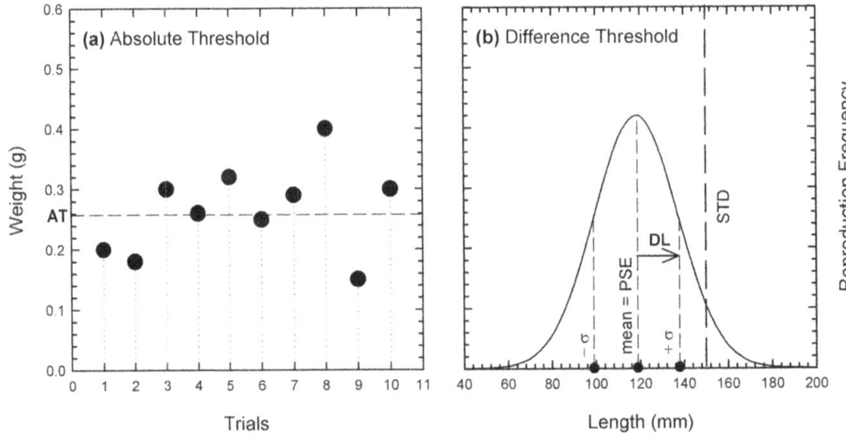

Figure 5.1 Examples of determining thresholds with the method of adjustment.

In Figure 5.1 theoretical examples of determining absolute (a) and difference (b) thresholds are shown using the method of adjustment or reproduction. In (a), each trial involves adjustment of the weight of a probe which is placed on a participant's palm until it is sensed. This process is repeated N times (in this example $N = 10$) and the average (marked by the horizontal) is taken to be the absolute threshold. In (b), participants are asked to reproduce a 140 mm length (STD) by indicating the starting and end points with their finger. The mean of the normal distribution of the reproductions is taken to be the PSE, and the standard deviation the difference threshold.

Constant Stimuli

The method of constant stimuli requires the participant to identify the presence of a stimulus or the difference between two stimuli. In the case of absolute threshold, the participant is presented with a stimulus and then asked to report whether the stimulus was present or not by responding 'yes' or 'no' (this is known as the 'yes–no' paradigm), respectively. In each trial, the intensity of the stimulus varies randomly within a range which contains a limited number of intensities. Each intensity is tested several times and at the end of the experiment the proportion of 'yes' responses for each intensity is plotted against the corresponding intensity. In principle, the individual performance can be described by a sigmoid *psychometric function*, which approaches zero at lower intensities (i.e. low probability of detection) and one at higher intensities (i.e. higher probability of detection). For example, if the experimenter is interested in the passive perception of weight applied on the skin, they may present the participants with a number of weights and then ask them to respond with a 'yes' if they felt the weight or 'no' if they did not. The proportion of 'yes' responses for each weight is plotted against the weight intensity and a psychometric function is fitted to the data. The absolute threshold is the weight corresponding to 50 per cent of the psychometric function (Figure 5.2a). In the case of difference threshold, participants explore a standard and a test stimulus that are part of a range of stimuli. Usually, the range contains 7 or 9 stimuli with the standard stimulus being the mid range. The two stimuli could be separated in time (e.g. STD is presented first and test second, or vice versa), space (e.g. STD and test stimuli are presented simultaneously but in different spatial locations), or space and time (e.g. STD and test are presented in different times and different spatial locations). Then participants are asked to respond which stimulus was greater in magnitude (e.g. 'heavier' in the case of weight discrimination, 'longer' in the case of length discrimination). At the end of the experiment, individual data of the probability of 'heavier' responses are plotted against the stimulus intensities used and a psychometric function is fitted. The difference threshold, or JND, is taken to be half the difference between the thresholds at probability 0.75 (above which

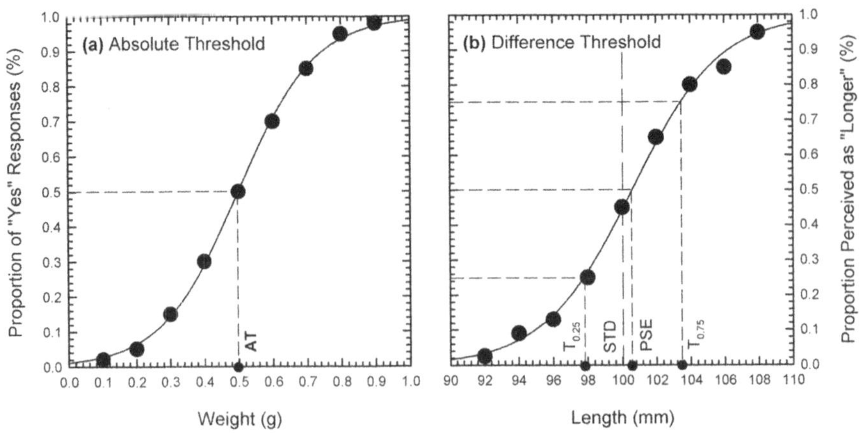

Figure 5.2 Examples of obtaining thresholds with constant stimuli.

'heavier' responses are very frequent) and 0.25 (below which 'heavier' responses are very infrequent) – that is, JND = $(T_{0.75} - T_{0.25})/2$. This paradigm requires a lot of trials per stimulus and no active part by the participant. Potentially, this can be a very accurate method of measuring absolute and difference thresholds but it requires a great number of repetitions and a high level of attention.

In Figure 5.2 theoretical examples of obtaining absolute (a) and difference (b) thresholds with the method of constant stimuli are shown. In (a), each point represents the frequency of 'yes' responses to the presence of a weight applied on the skin. Near zero, where the weight is extremely light, the majority of responses is 'no', since it is very difficult to detect the weight. However, as the weight is getting heavier, the number of 'yes' responses increases since it is easier to detect the weight. The absolute threshold is the weight where 50 per cent of 'yes' responses are recorded. In (b), each point represents the frequency of 'longer' responses to the presence of a length that is physically shorter or longer than the standard length of 100mm.

Staircase

The staircase method, which is also known as the 'up-and-down' method (Cornsweet 1962; Levitt 1970), is based on the idea of approaching the threshold on correct answers and receding from it on wrong answers. In principle, the up-and-down pattern of the participant's responses will, eventually, converge around the true threshold. Each experimental session starts with a stimulus intensity which is much stronger or weaker than the threshold in order to increase the chance of a correct response. Every time there is a correct response, the intensity of the stimulus is reduced (or increased) by a certain amount, called the *step size*. As the stimulus intensity approaches the threshold, the probability of a correct response occurring decreases and eventually the participant will give a wrong response. When the responses change from correct to wrong (and vice versa) a *reversal* has occurred. This process continues until a predefined number of reversals have occurred. In the end, the threshold is estimated on the basis of a certain number of reversals (usually between 6 and 10). In the case of absolute threshold, responses should oscillate around an intensity that allows the participant to reliably detect the existence of the stimulus. For example, in the case of passive weight perception, the intensity could start at near zero, increasing when participants fails to detect the weight applied on their skin and decreasing when they detect the weight. Eventually, a participant's responses will oscillate around an intensity which is the absolute threshold (Figure 5.3a). In the case of length discrimination, the participant is asked to discriminate between a standard length and longer or shorter test lengths. In the beginning of the experimental session, test lengths are much longer or shorter than the standard length to allow easy discrimination. Every time there is a correct discrimination, the distance between the standard and test lengths reduces by a given amount until participants give a wrong response which will increase the distance between standard and test stimuli. This process continues until correct and wrong responses

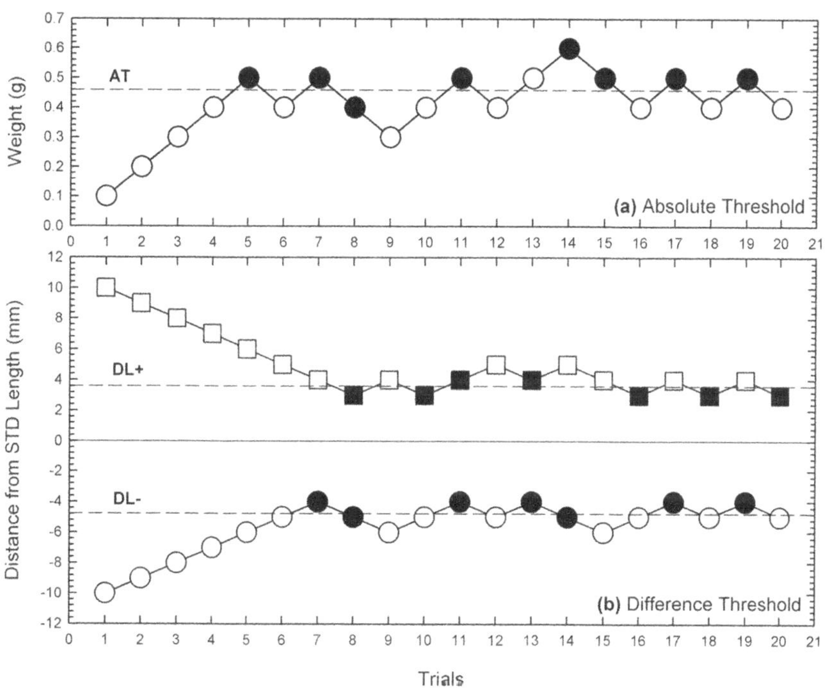

Figure 5.3 Examples of determining thresholds using the staircase method.

oscillate around intensities which approximate the difference thresholds for longer and shorter lengths (Figure 5.3b).

Figure 5.3 shows theoretical examples of absolute (a) and difference (b) thresholds obtained using the staircase method. In (a), white discs indicate detected stimuli and black discs undetected stimuli. The stimulus weight starts at very low value (0.1 g) and increases, or decreases, by a step size of 0.1 g after a negative ('no stimulus present'), or positive 'stimulus present' answer, respectively. Eventually, the responses will oscillate around a value which will be a good approximation of the absolute threshold. In (b), white discs indicate 'correct' discrimination and black discs 'wrong' discrimination. The difference threshold is approached from two directions: shorter and longer than the standard length. When there is a 'correct' discrimination, longer and shorter test lengths decrease and increase, respectively. When there is a 'wrong' discrimination, longer and shorter test lengths increase and decrease, respectively. Eventually, responses for both test stimuli will oscillate around values that do not allow reliable discrimination from the standard length. Note that the absolute values of the difference threshold for longer (DL+) and shorter (DL–) lengths do not have to be equal.

Psychophysical methods can result in lengthy experiments but, when combined with the right apparatus and stimulus, they can provide a powerful tool to study the sense of touch. Over the years, researchers have tried to build apparatuses that produce effective stimulation which allow appropriate investigation of different aspects of touch. The introduction of haptic interfaces has substantially facilitated this process by offering the opportunity to generate stimuli that can be automatically integrated with computerized psychophysical procedures.

Haptic Interfaces

Haptic Stimulation

In order to conduct haptics experiments one needs to be able to generate controlled stimulation that can be explored effectively by touch. Before the existence of the modern haptic interfaces, haptics researchers had to rely exclusively on building their own apparatus and haptic stimuli using real materials (e.g. wood, metal, plastic). The advantage of a hand-made, purpose-built apparatus and stimulus is that it is affordable and allows the researcher to have complete control over the stimulus and the haptic variable that they wish to investigate. However, it can be very time consuming (and monotonous), particularly if the generation of a large number of intensities is required. In addition, data collection often cannot be fully automated due to the limited connectivity of the apparatus with a computer.

With the arrival of the modern haptic interfaces, researchers have the option of producing haptic stimulation faster and, in some cases, more precisely. The advantage of a good haptic interface is that it can generate a continuum of stimulus values instantly and on demand. In addition, it is designed to operate via a computer which makes it suitable for automated data collection. However, good haptic interfaces are expensive and require strong programming skills to produce effective stimuli. In addition, the type of stimulation is limited primarily to proprioception, therefore ignoring a large part of haptic feedback coming from skin mechanoreceptors.

The most commonly used haptic interfaces work on the principle of generating geometrical and force properties of the interaction between the haptic system and physical objects. When we explore an object we perceive its shape and size on the basis of sensing normal forces at specific geometric locations. For example, the percept of a cube is formed because we experience normal forces exerted by the six surfaces (see Figure 5.4). Therefore, a haptic device produces a cube sensation by reproducing these forces at specific spatial locations.

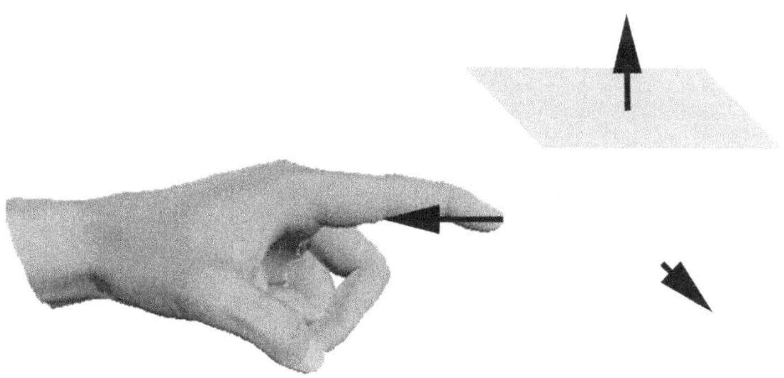

Figure 5.4 Forces experienced in the exploration of physical objects.

Popular Haptic Interfaces

Some of the most popular haptic interfaces are force feedback devices and are based on a robot-arm and delta design (Figure 5.5). They use a single-point contact and some have 6DOF (three translational and three rotational) which allow force and position sensing in *X*, *Y* and *Z* directions. The contact between the user and the virtual object is achieved through a probe, which could be a thimble, a stylus or a handle. The devices come in different sizes and motors and, in principle, the larger the device the larger the interactive space and the greater the forces they use. Therefore, depending on its size, each device allows the involvement of a different part of the haptic system. For example, devices with small workspaces up to 270 mm^3 (e.g. PHANToM® 1.5, Falcon, Omega) generate simulations that can be

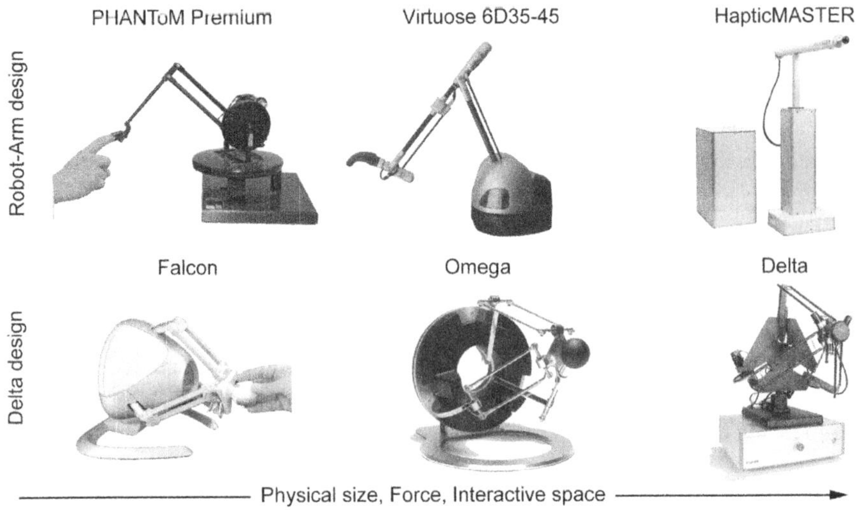

Figure 5.5 Some commonly used haptic interfaces.

explored fully only by the wrist joint, while devices with larger workspaces above 330 mm^3 (e.g. PHANToM® 3.0, HapticMaster, Delta) generate simulations that can also involve significant elbow and shoulder joint movements.

The single-point contact nature of these devices limits their ability to engage the haptic system in a natural manner. When we explore the environment with our touch we use multi-point contact in order to handle objects more effectively and maximize information gain. Proprioceptive information from more than one finger joint can give us instant information about geometrical properties such as size and shape. It has been found that simultaneous integration of spatial information from four fingers during exploration results in better shape recognition than consecutive integration of spatial information coming from just two fingers (Davidson 1972; Davidson and Whitson 1974). There are purpose-built multi-point contact devices including the MasterFinger and the CyberGrasp™. The MasterFinger is a desktop two-point contact haptic interface which allows position and force feedback through two miniature robot-arms. In addition, MasterFinger provides a thimble probe that can measure normal and parallel forces exerted by the fingertip during contact with the virtual surface (Oyarzabal et al. 2007). Moreover, in principle, a twin MasterFinger setup has the potential of generating stimulation for bimanual manipulation using precision grip. However, MasterFinger is under development (see http://www.immersence.info) and has still to be evaluated with psychophysical techniques. The CyberGrasp™ is a haptic interface which provides grasp-force feedback to each finger with the aid of five actuators. It is based on an exoskeleton design and in combination with CyberGlove™, CyberTouch™ and CyberForce™ can generate a wide range of haptic stimulation with visual feedback (http://www.immersion.com). For example, CyberGlove™ provides visual feedback of the movement of the hand. It can deliver vibrotactile stimulation to each fingertip which can be used to code physical intensities and is a large-scale force-feedback device that can generate stimulation for the elbow and shoulder joints. Even though it is capable of delivering a range of tactile stimulation, the Cyber-System requires the user to wear the exoskeleton which, together with the cables, could restrict natural movement. It is also very expensive which makes it less accessible to a wider range of users including haptics researchers.

It is also possible to design a multi-point contact haptic interface by combining more than one single-point contact device, which generally seems to improve haptic simulation (Wall and Harwin 2001; Michelitsch, van Veen and van Erp 2002; McKnight et al. 2004). However, these designs reduce the workspace dramatically and therefore further restrict the movements of active touch (Figure 5.6). Moreover, the limited point contact that these devices offer may also affect the way we perceive the weight of virtual objects. Research has shown that force distribution between the fingers differs during weight lifting (Kinoshita, Kawai and Ikuta 1995; Kinoshita, Murade and Bandu 1996), and this could affect weight perception (Flanagan and Bandomir 2000; Giachritsis and Wing 2007). Furthermore, weight perception within a virtual setup may also be affected by the inherited lack of cutaneous feedback. Cutaneous feedback does not only allow us to perceive non-geometrical object qualities, such as material, texture and temperature, but it is also used to initiate grip forces which have been found to affect weight perception (Flanagan et al. 1995).

Despite their limitations, single-point contact devices can be used effectively to investigate the way we perceive certain geometrical aspects of physical objects such

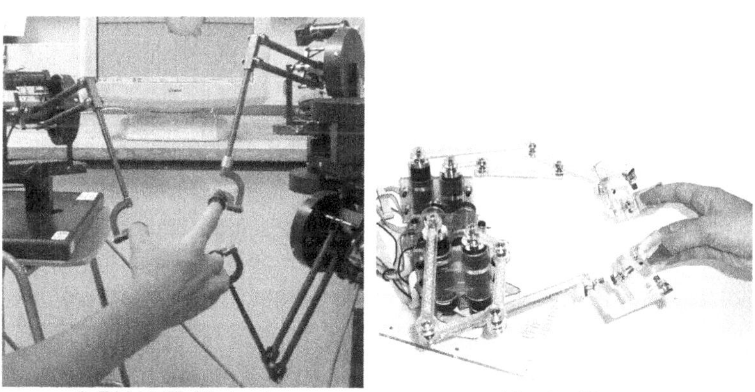

Figure 5.6 Multi-point contact haptic interface designs. *Note*: These can be achieved either by combining more than one single-point contact interfaces, such as PHANToM® 1.5 (left), or by purpose built systems such as MasterFinger (right).

as size, shape and orientation. The next section presents an example of psychophysical investigation of the perception of surface orientation using a PHANToM® interface.

Examples of Using Haptic Interfaces in Haptics Research: The Role of Spatial Integration in Tactile Slant Perception

Background

Perception of surface orientation is a fundamental ability of the haptic system since it is involved in both shape perception and object manipulation. In vision, it is known that perception of line orientation in the human fovea improves with length. Discrimination judgements of the orientation of a linear segment, relative to horizontal, improve with length and reach an optimum when the line is about 0.5° long (Westheimer and Ley 1997). This suggests that integration of spatial information may be used to improve orientation judgments. In order to find out if similar principles underlie perception of orientation of a haptic line, Giachritsis, Wing and Lovell (2006) conducted a psychophysical study using the methods of reproduction and staircase. The hypothesis was that if spatial integration was used to perceive surface orientation then the longer the surface the more accurate the percept should be. However, if spatial integration was not used to perceive surface orientation then the accuracy of the orientation percept should be independent of surface length.

Methods

Apparatus and Stimuli The reproduction and staircase methods were used to measure difference thresholds for orientation perception. Four haptic lines with lengths of 35, 70, 140 and 245 mm were generated using a PHANToM® Premium 1.5 interface. Their orientation varied in the radial direction between 0° and 90°.

Set Up and Procedures The contact between the participants and the virtual lines was achieved through a thimble which was attached at the end of the PHANToM arm. Participants placed the fingertip of their left index finger in the thimble and explored the virtual line for 3 seconds by moving smoothly back and forth along its entire length. They had their eyes closed for the whole duration of the experiment.

The reproduction procedure included two phases. In the first phase, the participants explored the virtual line in the radial direction and in the second phase they reproduced its orientation by indicating the bottom and top ends of the line with the help of a foot operated switch. Participants reproduced ten random orientations for each length. The staircase procedure included two staircases which ran simultaneously. The first staircase approached the difference threshold from above the reference orientation and the second from below. Each trial included three phases: two exploration phases and a response phase. In the first and second phases, participants explored two lines of equal lengths but different orientations (the line was either steeper or shallower than the first one, depending on the staircase). In the third phase, participants indicated which line felt steeper (Figure 5.7).

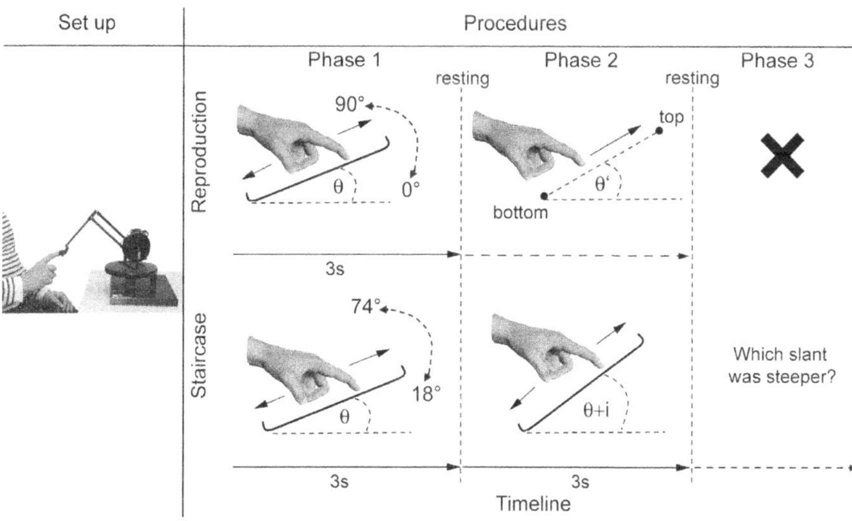

Figure 5.7 Experimental set-up and procedures.

Results and Discussion

Results were summarized by plotting the absolute values for the difference thresholds against the four lengths. Both methods produce very similar results showing that reproduction and discrimination of haptic orientation improve with surface length. Both difference thresholds decrease with increasing line length (Figure 5.8). Moreover, this improvement in judging haptic orientation with longer lengths follows an exponential decay pattern with the fastest rate of improvement occurring between 35 and 140 mm. Beyond 140 mm performance seemed to improve only slightly.

These results show that the haptic system may integrate spatial information to reach safer decisions about geometrical properties of surfaces, such as orientation. Even though, in principle, two points suffice to define a unique orientation, the haptic system seems to take advantage of redundant information in order to increase the reliability of its percept.

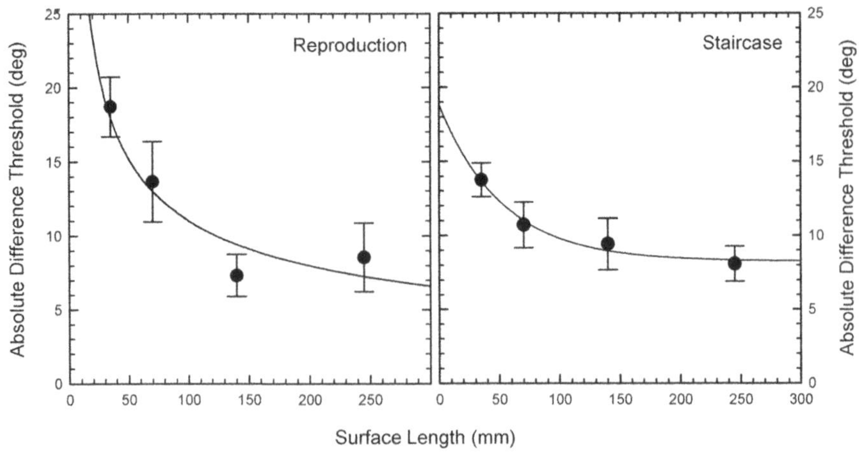

Figure 5.8 Overall performance of six participants in reproducing and discriminating surface orientation in the radial direction. *Note*: Both difference thresholds decrease with increasing surface length.

Conclusions

This chapter introduced central concepts in haptics research and fundamental psychophysical techniques that are used to investigate how the haptic system works. Recent advances in haptic interfaces have made it possible to create an integrated computerized tool which offers fully automated stimulus presentation, data collection and analysis. However, the most commonly used haptic interfaces cannot generate high-fidelity haptic stimulation since they often use one point contact, while simulation of fundamental physical properties such as texture, temperature and weight is either absent or very basic. Nonetheless, even though they under-stimulate the sense of touch, they can still be used reliably to investigate certain aspects of the haptic system.

Acknowledgements

This work was funded by the European Commission under the IMMERSENCE Integrated Project of the Sixth Frame Program (IST-4–027141-IP).

The author would like to thank the following companies, organizations and individuals for providing pictures of haptic interfaces:

Virtuose 6D35–45. Courtesy of HAPTION. Copyright Haption 2008.
HapticMaster. Courtesy of Virtalis. Copyright Virtalis 2008.
Falcon. Courtesy of Novint. Copyright Novint Technologies, Inc. 2008.

Omega and Delta. Courtesy of Force Dimension, Copyright Force Dimension 2008.
MasterFinger, Courtesy of Manuel Ferre, Universidad Politécnica de Madrid.
3 PHANToM design. Courtesy of William Harwin and Nick Melder, University of Reading

Note

1. While psychophysics focuses on how we perceive properties and events through tactile exploration, other methods such as *movement analysis* and *forces analysis* investigate the movement patterns and the forces involved. Patterns of reaching, exploratory and lifting movements, as well as grip and lifting forces, are also very important in understanding

References

Brisben, A.J., Hsiao, S.S. and Johnson, K.O. (1999), 'Detection of vibration transmitted through an object grasped in the hand', *Journal of Neurophysiology* 81: 1548–58.

Brodie, E.E and Ross, H.E. (1984), 'Sensorimotor mechanism in weight discrimination', *Perception and Psychophysics* 36(5): 477–81.

Cornsweet, T.N. (1962) 'The staircase-method in psychophysics', *American Journal of Psychology* 75: 485–91.

Davidson, P.W. (1972), 'Haptic judgments of curvature by blind and sighted humans', *Journal of Experimental Psychology* 93(1): 43–55.

Davidson, P.W. and Whitson, T.T. (1974), 'Haptic equivalence matching of curvature by blind and sighted humans', *Journal of Experimental Psychology* 102(4): 687–890.

Flanagan, J.R. and Bandomir, C.A. (2000), 'Coming to grips with weight perception: Effects of grasp configuration on weight heaviness', *Perception and Psychophysics* 62(6): 1204–19.

Flanagan, J.R., Wing, A.M., Allison, S. and Spenceley, A. (1995), 'Effects of surface texture on weight perception when lifting objects with a precision grip', *Perception and Psychophysics* 57(3): 282–90.

Gescheider, G.A. (1997), *Psychophysics: The Fundamentals*, Hillsdale, NJ: Lawrence Erlbaum.

Giachritsis, C.D. and Wing, A.M. (2007), 'Unimanual and bimanual weight judgments', in Proceedings of EPS Conference 2007, University of Edinburgh, UK, 4–7 July, Abstract, p. 83.

Giachritsis, C.D., Wing, A.M. and Lovell, P.G. (2006), 'The role of spatial integration in the perception of surface orientation with active touch', in Proceedings of EPS Conference 2006, University of Birmingham, UK, 10–12 April, Abstract, p. 72.

Gibson, J.J. (1962), 'Observations on active touch', *Psychological Review* 69(6): 477–91.

Johnson, K.O. and Hsiao, S.S. (1992), 'Neural mechanisms of tactual form and texture perception', *Annual Reviews in Neuroscience* 15: 227–50.

Kinoshita, H., Kawai, S. and Ikuta, K. (1995), 'Contributions and co-ordination of individual fingers in multiple finger prehension', *Ergonomics* 38(6): 1212–30.

Kinoshita, H., Murade, T. and Bandu, T. (1996), 'Grip posture and forces during holding cylindrical objects with circular grips', *Ergonomics* 39(9): 1163–76.

Levitt, H. (1970), 'Transformed up-down methods in psychophysics', *Journal of Acoustical Society of America* 49(2): 467–77.

McKnight, S., Smelder, N., Barrow, A.L., Harwin, W.S. and Wann, J.P. (2004), 'Psychophysical size discrimination using multi-fingered haptic interfaces', in Proceedings of Eurohaptics 2004, Technische Universität München, Germany, 8–10 June, pp. 274–81.

Michelitsch, G., van Veen, H.A.H.C. and van Erp, J.B.F. (2002), 'Multi-finger haptic interaction within the MIAMM Project', in Proceedings of Eurohaptics 2002, University of Edinburgh, UK, 8–10 July, pp. 144–9.

Oyarzabal, M., Ferre, M., Cobos, S., Monroy, M. Barrio, J. and Ortego, J. (2007), 'Multi-finger haptic interface for collaborative tasks in virtual environments', in J. Jacko (ed.), *Human-Computer Interaction: Interaction Platforms and Techniques*, Part II, LNSC 4551, Proceedings of 12th International Conference, HCI International 2007, Beijing, 22–27 July, pp. 673–80.

Ross, H.E. and Brodie, E.E. (1987), 'Weber's fractions for weight and mass as a function of stimulus intensity', *Quarterly Journal of Experimental Psychology* 39(A): 77–88.

Schiffman, H.R (2000), *Sensation and Perception*, New York: Wiley.

Verillo, R.T., Bolanowski, S.J. and McGlone, F.P. (1999), 'Subjective magnitude of tactile rougness', *Somatosensory and Motor Research* 16(4): 352–60.

Wall, S.A. and Harwin, W.S. (2001), 'Design of a multiple contact point haptic interface', in Proceedings of Eurohaptics 2001, University of Birmingham, UK, 1–4 July, pp. 146–8.

Westheimer, G. and Ley, E.J. (1997), 'Spatial and temporal integration of signals in foveal line orientation', *Journal of Neurophysiology* 77: 2677–84.

Wetherill, C.B. and Levitt, H.(1965), 'Sequential estimation of points on a psychometric function', *British Journal of Mathematical and Statistical Psychology* 18(1): 1–10.

–6–

Tactual Explorations: A Tactile Interpretation of a Museum Exhibit through Tactile Art Works and Augmented Reality

Isil Onol

This chapter introduces *Tactual Explorations* as an exhibition concept and analyses the performance of its first research-based inclusive public event which took place in Huddersfield between 29 September and 8 October 2006. The event included workshops, talks and a tactile exhibition focusing on one museum object. At this exhibition, the selected object, the bronze bust of Sophocles from the British Museum's Greek and Roman Antiquities collection was represented tactually by ten artists, through one haptic simulation and twelve supporting art works to enhance the physical information. This chapter investigates the effect of a haptic display which is surrounded by physical art works representing the tactile properties of the museum object and discusses a new method of a tactile interpretation. A potential series of exhibitions is suggested, which are focused on individual museum pieces, and where visitors engage with a number of physical and virtual works in order to study the original. The chapter first examines the visitors' interaction with the exhibition as a whole, and then focuses on the results gained from the engagement with the haptic simulation alone. The comparison of both results is presented, not to replace the traditional museum display, but to enhance the information conveyed about the exhibit to a wider, more diverse audience.

Tactual Explorations forms part of current PhD research that is investigating the topic of haptic interactions with museum objects, under the supervision of Dr Steve Swindells from the University of Huddersfield and Ann Marie Shillito from Edinburgh College of Art.

Introduction: *Tactual Explorations* as Tactile Interpretation

Tactual Explorations is the title of research-based public events that focus on individual museum objects and create a tactile interpretation of them through several artists' virtual and physical works. These works are not simply inspired by but directly based on the selected object's tactile properties. As a pilot study, the first of these events took place at the Northlight Gallery in Huddersfield, UK, between 29 September and 8 October 2006. The event included relevant workshops, talks and a tactile exhibition that interpreted the bronze bust of Sophocles from the British Museum's Greek and Roman Antiquities collection. Ten commissioned artists represented this selected museum object tactually with one haptic simulation and twelve supporting art works to enhance the physical information available to the viewer (see Figure 6.1 and Table 6.1).

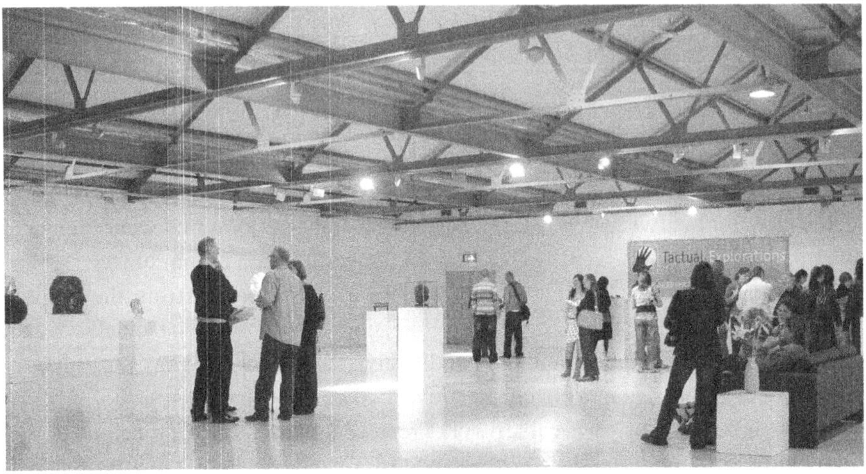

Figure 6.1 Tactual Explorations Exhibition at the Northlight Gallery, Huddersfield.

This project formed part of my practice-incorporated PhD research which is investigating the use of emerging (haptic) technologies and smart materials, their implementation into museum environments for object interpretation and also investigates solutions to create access to precious museum pieces without denying their conservation factors. The research outcomes are expected to serve as a foundation for a new interface, and its initial outcome will be the production of an interactive-haptic museum display that could be opened to the public.

For most curators, objects are the most important elements that can give museums their value and they would argue that without these artefacts, the museums would have no meaning. Thus, the presentation of these objects to the public is the actual starting point for *Tactual Explorations*, rather than acceptance or denial of current practices of museology.

Turning a 'natural object' into a 'humanly defined piece' is one way to explain object interpretation (Pearce 1994). A 'strong interpretation' has the potential to transform the meaning of the object completely by bringing out something that was not there in the first place (Carrier 2006), whereas inviting artists into the process

Table 6.1 Commissioned artists of the pilot study and their art works

Artist's Name	*Artwork Title(s)*	*Materials and Dimensions*
Tom Ainsworth	Expression	Cast silicone, black pigment and metal 14 cm × 10 cm × 2 cm
	Hair	Cast silicone, black pigment, metal and hair 14 cm × 10 cm × 2 cm
Carolyn Alexander	Unravelled	Latex 67 cm × 11 cm × 9 cm
Louise Atkinson	Hairball	Hair, hessian, cotton and stuffing 70 cm in diameter (approx.)
Lynn Cox	The Wiry Old Man	Wire 30 cm × 30 cm
Deborah Gardner	Viscid Head	Wax 75 cm × 63 cm × 43 cm
Isil Onol	Haptic Bust of Sophocles	Data on computer
	Surface	Brass 30 cm × 27 cm
Murat Ozkasim	Takes a Lot of Licking	Bronze and chocolate 8.5 cm × 7.5 cm × 6.5 cm
Megha Rajguru	Sophocles Circa 2000	Hair and plaster 39 cm × 22 cm
John Swindells	Inverted Head	RTV polyurethane 25 cm × 20 cm × 23 cm
Zoha Zokaie	Contours of the Face	Aluminium wire and paper, alginate, sand, three-part work, Each 30 cm × 35 cm × 30 cm
	The Tale of its Touch	Copper sheet 40 cm × 60 cm

can enhance the way surface information is conveyed rather than reinterpreting the meaning. *Tactual Explorations* aims to use artistic approaches to decipher visual information and to present a tactile interpretation of a museum object – not to replace it, but to enhance the existing information for all visitors regardless of their background or needs.

Objectives: Inclusive Design and Accessibility

Visually impaired museum-goers require better and wider access to museum collections and, now that government legislation officially recognizes this need (Candlin 2004), it is even more necessary for researchers to work towards better, more inclusive and more meaningful access to the museum exhibits for such visitors.

The concept of inclusive design is very important to my research and its practice. Tactile interpretation in museums has the potential to revolutionize the museum experience and increase access to the object for everyone, including physically impaired visitors. One of the first problems that we encounter in common 'inclusive' solutions in the design sector is that they tend to exclude part of their audience in order to include another part. A solution that creates further problems can still be considered as a solution, however not an ideal one. For instance, creating touch-only museum interpretations would only address a small number of visitors; this would therefore be against the idea of inclusive design (Pearson 2003: 41). Even though *Tactual Explorations* emphasizes touch as the main sense, the tactile exhibition of *Tactual Explorations* cannot be classified as touch-only. All exhibits can also work as stand-alone visual art works, vision on its own, however, does not define any of them. At all times, visitors are invited and encouraged to bring other senses into their experience to establish a multisensory interaction (Figure 6.2).

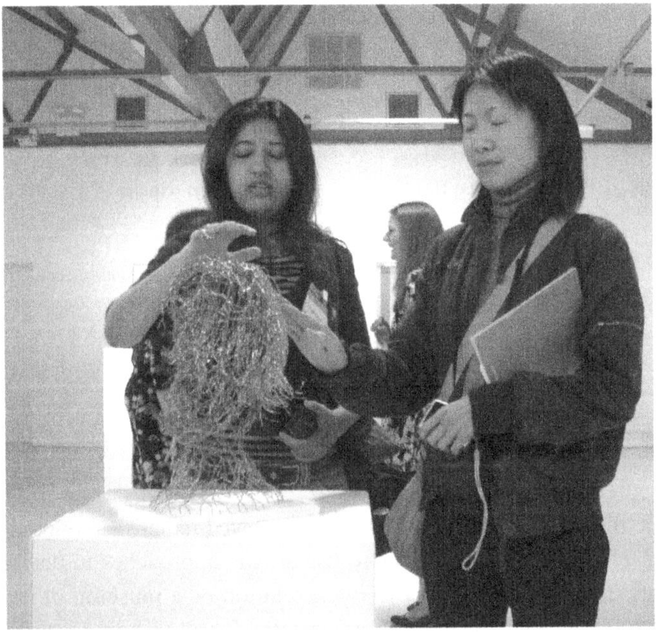

Figure 6.2 Visitors examining Lynn Cox's artwork 'The Wiry Old Man'.

Results of previous user feedback exercises have indicated that the addition of tactile feedback as a separate interface tool to a visual display can enhance the learning experience and increase the accuracy of the tactual perception. At the same time, the freedom of movement and the use of a tactile interface can create the illusion of one-to-one interaction with the original precious museum object. These exercises, however, were set in a controlled research environment and did not provide enough variety to test how inclusive and accessible an exhibit could be. In order to collect the necessary data for the research, it became crucial to create a public event that was set in real-life conditions. More precisely, a museum object was needed that could be examined by random visitors – not necessarily with the questions of this research in mind – if realistic outcomes were to be achieved.

In order to fulfil the requirements of an inclusive project from all points of view, it was important to create an event that was useful to people, not just to my research. The project proposal, funding and sponsorship applications, calls for volunteers for different roles, the artists' brief and the publicity materials were all produced with inclusivity aspects in mind. Commissioned artists were selected according to their research background, use of accessibility as concept and practice in their previous work and their approach to the artists' brief on their proposals. The curation of the exhibition involved paying a great deal of attention to the accessible materials (Figure 6.3).

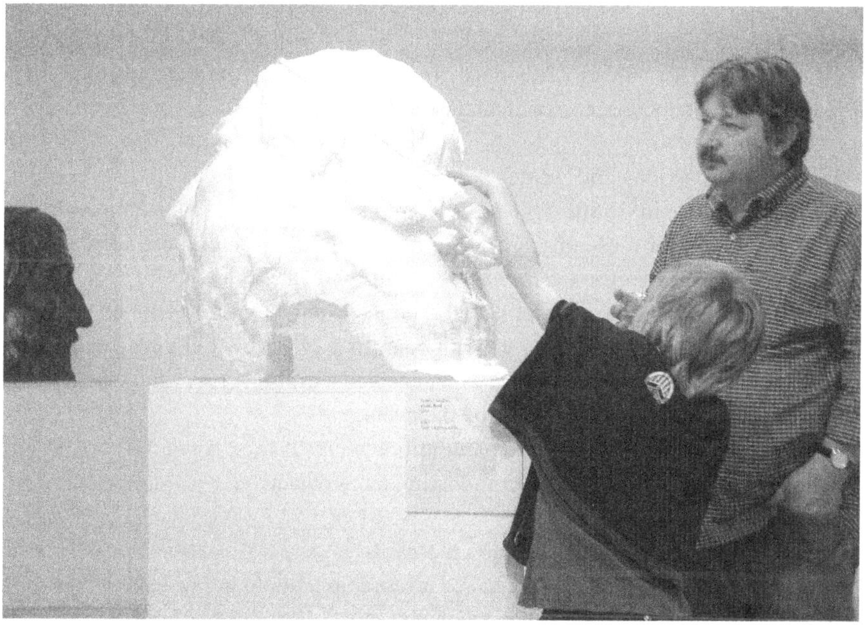

Figure 6.3 A young visitor examines Deborah Gardner's wax artwork 'Viscid Head'.

The gallery room allocated for this event was larger than that usually provided for exhibitions of this size and all physical barriers were removed. This allowed additional space for wheelchair-users and visitors with guide dogs and other aids to walk around the exhibits in comfort. Attendants were available at all times to rotate the exhibits or place them on a lower plinth temporarily if visitors required this.

Access to written information was provided in alternative formats for visitors who preferred or needed optional access. An audio guide was produced for a variety of platforms. This guide included extra help and an audio tour that covered all exhibits and artists' information. The audio guide first explained how everything was set up in the room (i.e. where on plinths the Braille labels could be found and how these labels corresponded to the objects discussed in the audio tour, etc.). Then it moved on to the four-part audio tour. The first part of this audio tour introduced the project; the second part discussed the individual art works; the third part provided short biographies of the artists; and the final part provided information about the original Sophocles bust in the British Museum. Listeners were able to skip parts, or re-listen at any point during the guide. A listening point with this audio guide was installed at the entrance. It was also available for visitors to carry with them in the following formats:

- As recorded tapes to use in old-style Walkman™ personal stereos that we marked with Braille labels (some visually impaired people prefer using this type of player with large buttons);
- As audio CDs to be used on the CD players we provided (or visitors' own players);
- As various digital files to download from our computers into the visitors' MP3 players or notebook/laptop computers.

All labels, instructions and written information, including the exhibition booklet, were also available in Braille. One of the commissioned artists, Lynn Cox, is also an access consultant for museums, and she provided Braille labelling for the art works and access consultancy for the *Tactual Explorations* event as a whole.

During the exhibition, special needs schools arranged group visits on different days. These groups included people with cognitive, physical, and visual or hearing impairments. Their attendance at the event enabled a truthful observation and testing of the access and facilities provided by the event.

The project and the pilot study intentions were to raise awareness of issues relating to 'inclusive exhibits' with following main objectives:

- To produce multisensory exhibition pieces;
- To use Augmented Reality to enhance information, not to impress the visitors;
- To provide an opportunity for people with limited or no sight to have access to art exhibits;

- To explore the main and hidden aspects of 'touch' in an object that is exhibited;
- To analyse and represent the tactile information (of an exhibition object) by layering the sensory properties involved and individually highlighting these properties;
- To show that accessible gallery experiences can be achieved on low budgets;
- To bring precious exhibits in the larger museums to visitors living at distance from these establishments.

Production and Design Process

A project as ambitious and large as this required extra funding and sponsorship for the first event to be realized. Metropolitan Works Creative Industries Centre in London, who agreed to become one of the sponsors of the event, supplied the expertise and allocated the use of their £60,000 Faro Laser ScanArm to scan the bust of Sophocles into a CAD program. The University of Huddersfield funded the artists' fees; Oxidise, a London-based design company, developed the website; many volunteers, who are experts in their subjects, offered their services free of charge, and funds were raised from public and private businesses.

For the exhibition, each artist created an art work or set of individual pieces to represent their chosen tactile information from the surface of Sophocles's bust according to the guidelines given in the artists' brief. By paying particular attention to accessibility, each piece was designed to be explored mainly (but not only) by touch. In short, the purpose of the exhibition (as an interpretation of a museum object) was to demonstrate how one museum piece could be explored tactually and what elements of the tactile perception could be represented by using a variety of materials. One of the objectives was to achieve a tactile setting in which, although each piece in the exhibition would speak for itself as an art work, when presented together in the exhibition room, the distinctive pieces would form one big tactile representation of a museum object.

Because the main aim was to explore the tactile information of a selected object through the artists' representations of the tactile information, the art work production techniques were not limited to the object's own. As a general rule of the *Tactual Explorations* concept, the artists were free to explore different materials in order to achieve the correct effect of their given/chosen interpretation of the tactile properties. Artists could focus on a detail taken from the original piece, or represent the whole piece, in order to form the complete haptic experience they had in mind. Each art work focused on one or more of the following properties acting as the main feature of the art work: vibration, surface texture, surface temperature, shape, slope, hardness, weight, elasticity, and pliability (McLinden and McCall 2002). In addition, the tactile properties of the artists' works were created to supplement the force-feedback of the haptic simulation (which only provides shape information, see Figure 6.4)

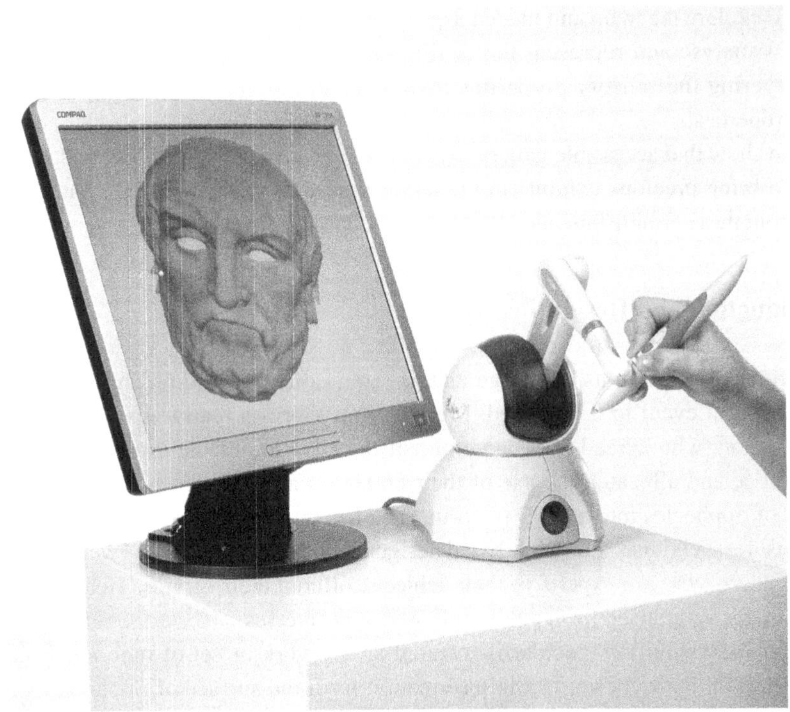

Figure 6.4 Haptic bust of the Sophocles bronze.

by adding the missing tactile feedback (which provides information about surface texture and material properties).

The sensation of touch was not limited to hands. Artist Murat Ozkasim noticed this point in the artists' brief and decided to create an art work that would encourage visitors to go one step further. With a tongue-in-cheek approach to Sophocles' form, Murat created edible replicas of this sculpture in order to represent an important – but often neglected – way to identify an object's shape by using the tactile receptors of the human tongue (Figure 6.5). To produce the casting of the chocolate sculptures, Murat first created a replica 6.5 cm high using a rapid prototyping method with technology supplied by 'WOW Academy' of Bradford, one of the sponsors of the event. The chocolate replicas were handed out to visitors.

Haptic Interaction

A good definition of the word 'haptic' – derived from the Greek *haptesthai* which means 'to touch' – is 'relating to or based on the sense of touch' (Merriam-Webster 2007). Haptic technology simulates the sense of touch using computers, creating

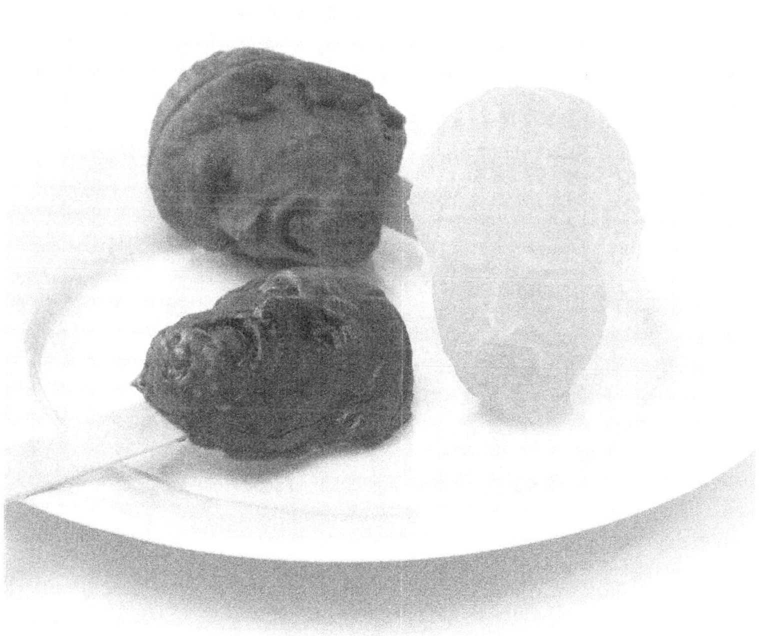

Figure 6.5 Murat Ozkasim's edible artwork, entitled 'Takes a lot of Licking', is shown with the 6.5-cm-high rapid-prototyped replica.

a two-way interaction with a device. The study of haptics has emerged from the combined use of robotics and computer graphics to create powerful visualization systems. Today, high performance force-feedback haptic displays are still relatively expensive to produce; however, medium-range ones are becoming more affordable and promise interesting possibilities for the future. As these devices are becoming more widely available, the software is getting easier to use and costs are coming down, and so it is now possible to design systems that introduce these technologies into everyday use. The device that was used on the first *Tactual Explorations* project is the PHANToM® Omni™, developed by SensAble Technologies.

At the *Tactual Explorations* exhibition, Haptics enabled artists and audiences from diverse backgrounds to come together to explore one famous museum object. Even though haptic technologies are relatively new and most of its applications became central to the event, at this exhibition it was used as a medium rather than being the focus of the exhibition. In fact, the physical art works completed the missing tactile elements. Haptic technologies and the physical objects of the exhibition almost existed to improve each other, as a whole augmented reality solution (Figure 6.6).

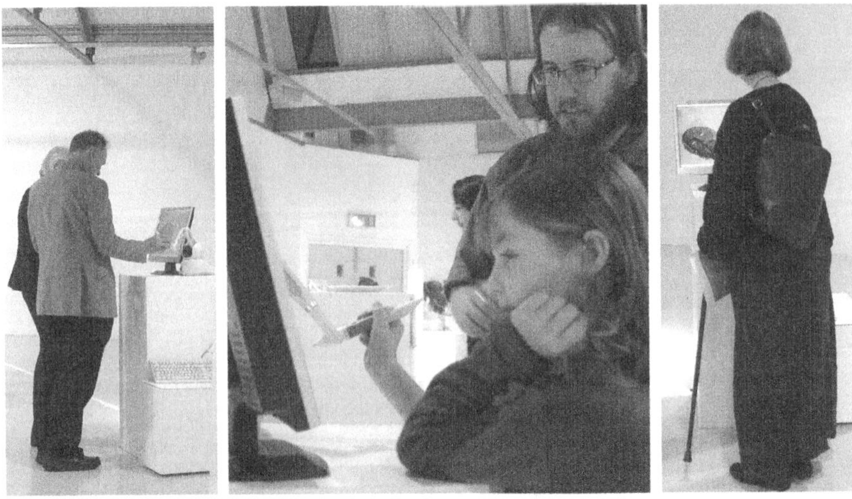

Figure 6.6 Visitors engaging with the haptic simulation.

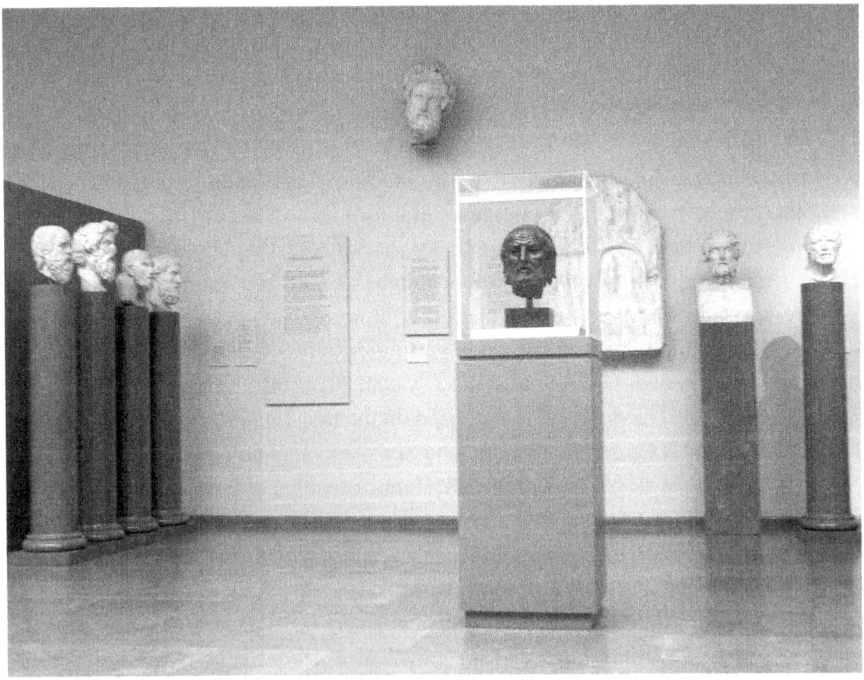

Figure 6.7 Bronze bust of Sophocles at the British Museum. Photograph by Isil Onol.

Selected Museum Object

The selected piece for the first *Tactual Explorations* exhibition (the pilot study), was the bronze bust of Sophocles, currently displayed in Room 22 of the Greek and Roman Antiquities section of the British Museum (see Figure 6.7).

Because the surface is too delicate to be handled regularly and original Greek bronze works on a large scale are quite rare, the bust is displayed in a glass case and touching is strictly forbidden. These restrictions, and the amount of surface detail available, however, make this exhibit a perfect object for the *Tactual Explorations* project. but it was, of course, necessary to purchase a replica of the bust for use at the exhibition and for the artists to examine physically (see Figure 6.8).

As Morgan (2006) says:

> Whilst [the object's] form is well rendered we can only imagine that, even if the bust weren't too delicate to touch, touching it would ultimately disappoint – the hair and beard offering the same cold and disinterested response to physical contact that the nose and lips, chin and cheeks would. A direct replica made available for visitors who are visually impaired would not necessarily make sense as the visitor may not be able to assign visual memories to surface details: how would one know Sophocles is wearing a headband if one could not see it?

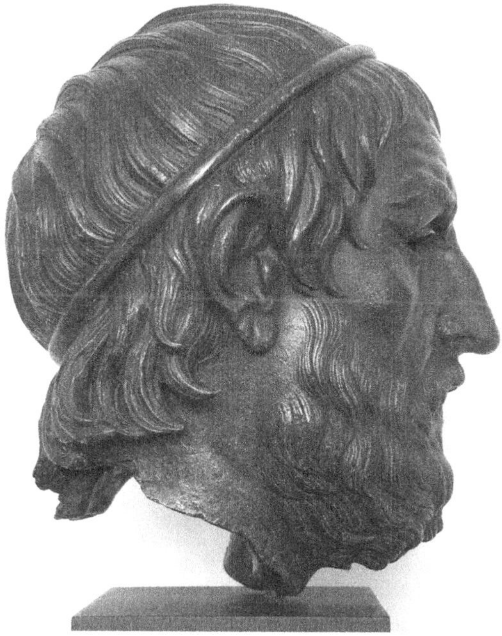

Figure 6.8 Replica bust of the bronze bust of Sophocles. Photograph by Isil Onol.

Workshops

As part of the *Tactual Explorations* exhibition, six workshops were held encouraging everyone, especially museum and gallery workers, school teachers, students and children, to explore some of the techniques behind the exhibits. The workshops also introduced tactile drawing methods, not only to visually impaired people, but to anyone interested in learning these methods. Most of the workshops were suitable for children and all workshops were free of charge. In total, 65 people attended these workshops. The workshop titles and their facilitators were:

- 'Drawing by touch', by Tom Ainsworth
- 'Access to art: Whose responsibility anyway?', by Caglar Kimyoncu
- 'Tactual drawing and mark making', by Carolyn Alexander
- '3D Collage', by Louise Atkinson (Figure 6.9)
- 'Sensory Stories', by Amy Hirst
- 'Tactile Drawing', by Lynn Cox (Figure 6.10)

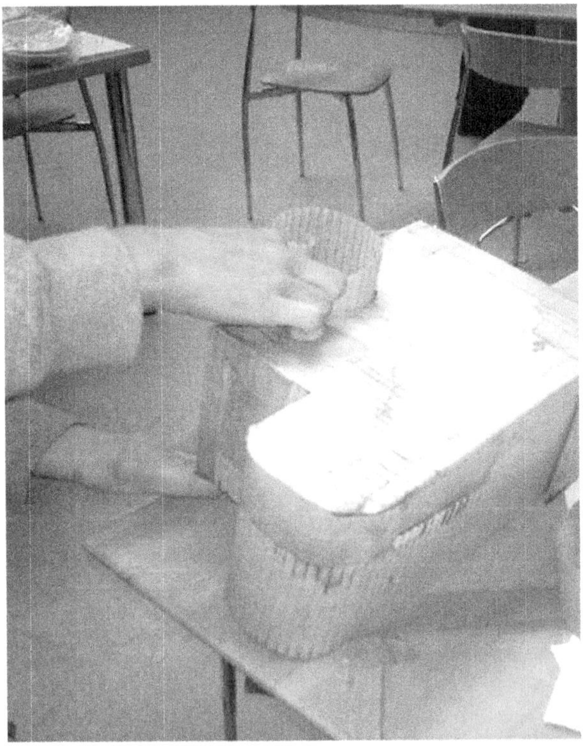

Figure 6.9 A participant creating work at Louise Atkinson's workshop '3D Collage'.

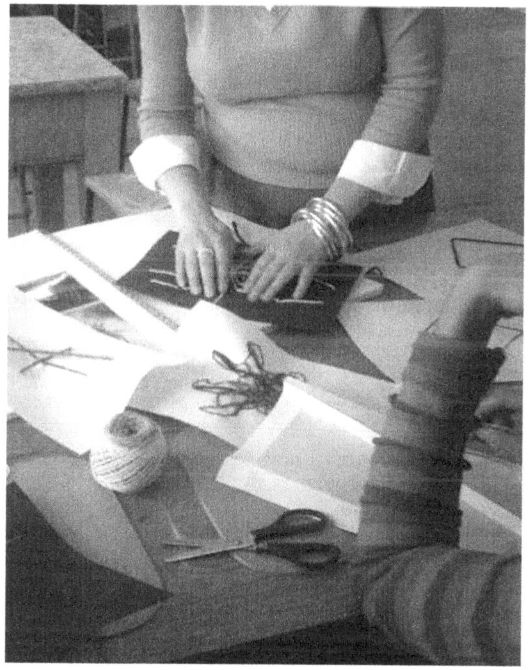

Figure 6.10 Lynn Cox instructing participants at her workshop 'Tactile Drawing'.

Results

The exhibition provided practical data vital to the progression of my practice-incorporated PhD and engaged audiences relevant to the research (e.g. visually impaired). Special needs schools in and around Huddersfield arranged visits to the exhibition and they gave valuable feedback on their experiences. More than 300 people attended the exhibition during the first five days and another sixty-five attended the workshops. The audience interaction with the art works was observed and recorded and feedback was obtained via questionnaires and a visitors' book.

The majority of the comments in the visitors' book and the feedback forms were related to 'being able to touch'. The only time people referred to the haptic simulation was when they were asked directly. It was important for this project to achieve a transparent use of technology and it was vital that technology did not take over but became merged with the rest of the exhibits.

For research purposes, the data enquiry first looked at the visitors' interaction with the exhibition as a whole and then focused on the results gained from the engagement with the haptic simulation alone. The aim was not to replace the traditional museum display, but to enhance the information conveyed about the exhibit, to a wider, more diverse audience.

Table 6.2 Experience feedback ratings from thirty randomly selected visitors

Experiences	*Feedback Ratings*				
	SD	*D*	*U*	*A*	*SA*
Tactile art works enhance the 'haptic' simulation by offering the missing physical information	0	0	0	22	8
Interpreting museum objects through tactile works is a very inclusive approach	0	0	0	21	9
With the help of the tactile art works, the computer interface seemed to vanish	0	0	5	12	13
After interacting with all the art works, I felt I interacted directly with the bust of Sophocles	0	0	2	10	18
I felt that the overall tactile interpretation provided space to add my own interpretation	0	0	0	2	28

Note: Key to responses: SD = strongly disagree; D = disagree; U = undecided, A = agree; SA = strongly agree

Five questions were asked to thirty randomly selected visitors, regardless of their disability, social status or cultural background (Table 6.2). These questions might be considered as 'leading' questions by many researchers, but in order to direct the participants to the use of the 'interface', rather than the technology itself, it was considered necessary to use some kind of guidance in the sentence structure.

Another thirty randomly selected visitors were asked more generic questions about attending museums and their feedback on the *Tactual Explorations* concept. According to their answers (figures rounded):

- 86 per cent had never been to the British Museum;
- 93 per cent visited the exhibition because of its tactile content;
- 93 per cent had never heard of the word 'haptic';
- 100 per cent had never engaged with a haptic device;
- 100 per cent would like to see other museum objects interpreted in a similar exhibition concept.

Among children and the younger audience, the average time spent with the haptic simulation was ten minutes and the average time spent in the exhibition room was thirty minutes. With adults, the time spent with haptic simulation was 6 minutes and the average time spent in the exhibition room was 25 minutes. In order to keep the visitors experience as genuine as possible, during the observation no questions were asked and therefore the results of this section were arranged as 'younger audience' and 'adults' purely from their appearance. The feedback forms included questions relating to visitors' age, occupation, and whether they considered themselves to have a disability or not. For the nature of this research, gender was not considered to be relevant and was therefore ignored.

There was also constructive criticism from the visitors. Even though the majority of the people (98 per cent) found the haptic device easy to use, the other 2 per cent expected to be provided with written instructions for using the device. It had not been considered appropriate to do this on the pilot study as it would have changed the results of the data analysis; however, it will be considered in future exhibitions. Some visitors suggested a blind-folded tour of the exhibition and a day allocated to adults only. These suggestions will also be taken into account. Most workshop attendees criticized the limited availability of the workshops; unfortunately, the budget did not allow any additional free workshops. For future events, repeat workshops will be provided in order to open up these activities to more people.

Conclusion

By investigating the effects of a haptic display surrounded by physical art works, the first event of *Tactual Explorations* attempted to offer a new method of interpretation for museum objects. The exhibition and its workshops were fully accessible and available to visitors free of charge in order to include everyone regardless of their financial or social status. The exhibition was designed especially, but not exclusively, for visually impaired visitors. At the time of commissioning artists, the competition was kept open to all artists regardless of their background and disabilities. The exhibition's ten selected artists came from diverse backgrounds.

A website has been launched providing details of the project and those involved, including the artists, volunteers and sponsors. This website was designed according to accessibility requirements by providing clear design, regular and high contrast versions and passing the XHTML/CSS tests (see www.tactual.org.uk).

A booklet in written format and Braille showing the work in progress was produced as reference material. These were handed out to visitors to the exhibition. Information in this booklet was also included in the audio guide.

The project made a difference by:

- Using augmented reality and haptic technology to enhance access to traditional art form;
- Providing opportunities for people with limited or no sight to have access to art exhibits;
- Bringing interpretation of precious exhibits in larger museums to visitors living away from these museums;
- Exploring the main and hidden aspects of 'touch' in respect of an object that is exhibited visually.

The following were available at the exhibition at all times:

- Braille labels on each exhibit;
- Audio guide (on cassettes, on listening points and available to download);
- Braille version of the exhibition booklet;
- Clear text on printed material and labels;
- Space around each exhibit for wheelchair users;
- Attendants to help visitors with anything from filling in forms to the use of equipment.

The next step for the *Tactual Explorations* concept is to create further exhibitions that interpret other museum objects by using the experience gained from the pilot study. As well as proposing these exhibitions for the museums where the selected object is located, it is also planned to take these exhibitions to people who live away from these institutions. The changes in the haptic and adaptive technologies and advancement in the research will also be incorporated into these events. Projects with small budgets will not sacrifice accessibility; priority will be given to elements that will make an event more inclusive and clearer. The results and papers of each *Tactual Explorations* event (including the pilot study) will be made available to museums and galleries as case studies for their future projects.

Acknowledgements

The author is grateful for the haptic programming provided by Alastair Barrow of the Interactive Systems Research Group, University of Reading. All photographs are by Murat Ozkasim, unless otherwise indicated.

References

Candlin, F. (2004), ‘Don’t touch! Hands off! Art, blindness and the conservation of expertise’, *Body Society* 10(1): 71–90.

Carrier, D. (2006), *Museum Skepticism: A History of the Display of Art in Public Galleries*, Durham, NC/London: Duke University Press.

McLinden, M. and McCall, S. (2002). *Learning through Touch: Supporting Children with Visual Impairment and Additional Difficulties*, London: David Fulton.

Morgan, B. (2006), *Tactual Explorations* Exhibition Booklet, Leeds: Footprint Workers Co-operative.

Merriam-Webster (2007). ‘Merriam-Webster’s Online Dictionary’, available online at http://www.m-w.com/dictionary (accessed 15 August 2007).

Pearce, S.M. (1994), ‘Museum objects’, in S.M. Pearce (ed.), *Interpreting Objects and Collections,* London/New York: Routledge, pp. 9–11.

Pearson, A. (2003), ‘A museum professional’, in E.S. Axel and N.S. Levent (eds), *Art Beyond Sight: A Resource Guide to Art, Creativity, and Visual Impairment*, New York: AFB Press, p. 503.

–7–

CONTACT: Digital Modelling of Object and Process in Artefact Teaching

Roger Doonan and *Michael Boyd*

The CONTACT project addresses identified problems in teaching material culture in Archaeology and Classics through creating real and virtual networks of students, expertise and material. The real networks bring students into contact with material through the mediation of experts in that material, and the virtual networks back up these experiences with virtual material and expertise tied to the students' real experiences. CONTACT is increasing flexibility in curriculum design within Archaeology and Classics, thus allowing material culture studies to assume a central position by removing restrictions associated with this area of curriculum design. By cementing existing networks of cooperation and expanding these through a programme of initiative funding, the development of an inclusive cooperative network of HE and related institutions is being established which is already resolving subject-related needs with respect to material culture studies.

A major part of the CONTACT process is a virtual learning environment (VLE) intended to reinforce and enhance students' experiences with real objects. To this end, the VLE models both the object of study and the process of analysing it. This paper presents both aspects of the VLE and examines the challenges of digitizing the sensory experience of an artefact as physical fact and as object of inquiry. An initial evaluation of the success of the project is presented on the basis of pilot studies with students.

Introduction

While assembling this chapter, one of us (RD) found ourselves sharing a bus journey with two teenagers eagerly discussing their choice of Christmas presents. The debate was passionate and increasingly vocal as the advantages of the Nintendo Wii were scored against the Xbox 360. The argument centred on whether the Xbox with its better graphics but conventional controllers amounted to a better package than the Wii with its admittedly poorer graphics but an interface which relied on bodily movement. It echoed some of the debates that we had been having earlier in the year concerning the digitization of material assemblages to use in teaching. Should we strive for the digital perfection? Should we satisfy ourselves with a model based loosely on process? Or should we in fact hold tight to process and render that as accurately as we could in the virtual world? As the teenagers began to use fists to support their points of view, it became clear why such debates are necessarily so impassioned. Just as the debate on the bus hadn't really been about graphics processors or game controllers, our heated debates at various committee meetings hadn't really been about digital resolution or process models, more they were about how we perceive and engage with the world – that is, how our senses are entwined in ways that we learn. This is something we feel has been underdeveloped in the field of pedagogy, especially in the context of using physical objects in teaching, and it is something we feel to be especially pertinent in light of the burgeoning numbers of projects that aim to virtualize the many objects that currently exist in collections. This chapter discusses some of the aspects of learning with objects and considers the implications of the current agendas in higher education which advocate e-learning and virtual learning environments. Although this chapter focuses on a case study within the domain of Archaeology and Classics, we suggest that there are valuable lessons which are relevant to any subject that deals with objects and materials.

In the recent benchmarking statement for Archaeology, the subject is defined as 'the study of the past through material remains' (Darvill et al. 2007) and artefacts form a key component of those material remains. Students are expected to be able to 'observe and describe different classes of primary archaeological data, and objectively record their characteristics', and teaching should include 'hands-on ... artefact handling and identification work' (ibid.). Similarly in the Classics benchmarking statement, archaeological material is seen as equal to literary and documentary sources as primary material for the subject, and the interpretation of material culture is seen as one of the key strands within Classics overall (Osborne et al. 2007).

This centrality of artefact studies should come as no surprise in subjects such as Archaeology and Classics. However, in recent years, growing concern has been expressed about material culture teaching in relation to increasing student numbers, staff time and access to sufficient collections and expertise. Haselgrove et al. (2001: 17) suggest that 'the supply of future specialists is a major problem' since 'material culture no longer figures strongly in undergraduate teaching and there are few MA courses in artefact studies', while Chitty notes a 'gradual dispersal of accumulated knowledge and experience in highly specialist areas of material culture studies' and:

> A shortage of trained specialist professionals and serious loss to research scholarship is being anticipated by some. The specific training needs of this part of the sector and of other vulnerable specialist groups merit attention. (Chitty 1999)

The CONTACT project, funded by HEFCE (Higher Education Funding Council for England), was created in response to these concerns. Our first step was to try more precisely to define the sector's concerns about material culture skills in graduates. The results of a survey (Andrews 2006) indicated an awareness of the issue and, importantly, a willingness to tackle it. The problems quantified in the CONTACT survey were already widely recognized amongst the community, as could be seen from respondents' remarks – for instance, 'we have made real efforts in the last two years [to prevent decline]'; and, 'there are now real initiatives to put material culture teaching back on the agenda'. It seems likely that the background to a thirty-year decline of material culture teaching is the move away from artefact-centred approaches at the heart of cultural historical Archaeology to scientific approaches in the 1970s and 1980s, in which 'objective' observations were made scientifically, and so intimate understanding, familiarity and connoisseurship became less important; and then the post-modern critiques of the discipline reinforced a move away from traditional object knowledge in the 1980s and 1990s (cf. Candlin, this volume). There seems to be a widespread acceptance within the discipline that despite the accepted validity of those critiques, the provision of core skills has been lost from curricula in both Archaeology and Classics.

This is not to suggest a return to traditional teaching methods, since not only have approaches to objects developed significantly in this time but also the context in which our subjects are taught has changed dramatically.

Among the most significant issues to have arisen in teaching artefact studies is the issue of student numbers and related issues of space provision. Whereas small group teaching was once the norm it is now the exception, and where it is still undertaken it often involves repeat teaching with concomitant wear on collections. While different departments have coped in different ways, there was general agreement that material culture is best taught through small groups, meaning that pressure on resources was understood as a real problem. One widespread strategy is to limit the units involving material culture teaching through capped option choices that are taken in the intermediate and higher levels of a degree programme. Other strategies included restricting artefact handling to small-group seminar work. It was clear from the survey that there was wide variation throughout the sector, from those who felt they were coping successfully to those who felt they were barely coping at all; one respondent noted, 'we do most of the teaching without any contact with real artefacts, students only read about them and look at pictures'.

A second core issue is accessing expertise and collections. All departments have some access to collections, but again our survey found wide variation throughout the sector. Many departments rely on private staff collections or on material temporarily available for research reasons, while access to external collections can be difficult and expensive to arrange. In terms of expertise, where this is not available in-house, a clear difficulty in arranging for external experts is noted. One response was 'there are very few younger specialists coming into the field and even fewer are in a position to offer teaching/training to students; the old ones are getting older, and they are only available because they have retired; they won't last forever'.

The reformed nature of research structures was highlighted as a third core issue by many respondents. The three-year PhD is widely seen as inadequate for in-depth immersion in a particular material culture category; one respondent noted:

> When [students] go on to postgraduate study, they cannot hope to acquire sufficient expertise in a particular category of material in a year, or even two, to be able then to complete a PhD in only three years. Seven years is what it used to take, if you were lucky; most people took ten.

This problem is compounded by a lack of provision at taught Masters level. A quarter of respondents stated that there was no material culture component to their Masters programmes and of those programmes that were focused on material culture, as little as 20 per cent of time was spent on practically based sessions involving material culture collections.

The CONTACT survey highlights issues of deep concern for the sector, but also shows that in the light of the benchmarking exercise, departments have been keen to address these issues, and are experiencing a measure of success in doing so. The eagerness with which museums and other institutions with a curatorial role have digitized their collections under the banner of access, coupled with the generally positive adoption of e-learning and student-centred learning, has meant that the virtual rendering of collections has come to be seen as an obvious and natural solution to many of the issues outlined above. The CONTACT project is the result of successive projects developed over an eight-year span and builds on a sequence of evaluations and trials of integrating electronic resources in object-centred teaching. Throughout this extended development we have always been cautious of substituting virtual experiences for real ones. Evaluation of how students learn from objects has enabled us to develop a deep understanding of the problems associated with rendering the rich sensory experiences of the real world in a virtual environment. Despite the advances in haptic interfaces (e.g. Giachritsis, this volume) and improvements in digital representations, we feel that much of the enthusiasm to 'virtualize' objects is misplaced as it is guided by inappropriate understandings of perception and learning theory. The role of tactile encounter and a critical evaluation of perception in the learning environment form a central strand in our analysis of changes in learning practice.

Tackling the Problem

The aim of the project is to enhance the role of material culture studies in Archaeology and Classics by addressing practical, intellectual and resource-related problems in curriculum design. In particular, CONTACT aims to offer departments and individuals a framework within which they can develop their own solutions, drawing on internal, external and CONTACT-mediated resources, answering for themselves questions such as:

- How do we tackle increased student numbers in approaching material culture studies?
- How do we ensure students come into contact with appropriate expertise and collections?
- Where should material culture studies be situated in a modern, theoretically literate curriculum?

CONTACT's core framework involves creating real and virtual networks of students, expertise and material. These networks facilitate the mobilization of these three resource types in new pedagogic arenas. Specifically, the project offers initiative funding to individuals within departments to allow them to create these real and virtual networks.

What do we mean by 'real network'? We use 'real network' to describe 'traditional' teaching of students in class by experts using real material. The network is created by identifying these resources and bringing them together within the CONTACT framework. We have been working with collaborators in identifying suitable collections and material culture experts to carry out the teaching. The real network is central to the project as it is only through tactile engagement that objects come to be known.

Our first pilot project, at the University of York, was based around a first-year field Archaeology module which introduces students to a range of field techniques in conjunction with a departmental research project. Previously, resources to support student analysis and identification of artefacts had been limited. In 2006 a small group of students were able to enhance their understanding of this experience through an introductory lecture and hands-on session with a ceramics specialist, followed by a second hands-on session in Hull Museum. In 2007, the sessions were timetabled to coincide with the artefact sorting. The hands-on session was redesigned using a series of assemblages of different site types and ceramic dates. A prompt sheet was produced to lead students through the process of analysis, and reference cards produced for each period to provide students with a guide to the major ware types within the region. The major difference in 2007 was the creation of resources which encouraged students to develop their own (individual and group) skills of systematic analysis and interpretation, without relying on the specialist for advice. CONTACT is now embedded in the curriculum at York, and the resources produced are available for use by others.

On the face of it, the success experienced at York in resourcing and implementing the real network should beg the question as to whether a virtual network is necessary at all. However, the York pilot has been run with a test group of students, and in rolling it out to the entire cohort the key question of student numbers is encountered.

In CONTACT, the virtual network is conceived of as a strategy to *augment*, not replace, real teaching, and hence offer one way of dealing with the problem of teaching hands-on material culture studies in large groups. This augmentation comes about by allowing the students access to the materials in their own time and at their own pace, in addition to the timetabled window. Projects utilize the virtual network in a tightly defined curriculum where it supplements and builds on the real teaching at the core of the project. Beyond this, the virtual network offers several possibilities. It supports considerable flexibility of pedagogic pathways through collections whereby institutions can incorporate the features of the CONTACT collections within their particular curriculum design. Such pathways can connect the study of material culture to the wider curriculum, including critical analysis of objects, scientific and contextual analyses, evaluation of collections, and conservation issues.

Touch, Sight and the CONTACT VLE

As mentioned in the introduction, the digital representation of artefacts has burgeoned in the last decade. The advantages and disadvantages of the 'digital avatar' are discussed at length by Renaud (2002). He notes that the digitized artefact is a model that lacks *dasein* – an existence in the real world – but makes up for this through aesthetic representation and being open to experimentation and manipulation. Whilst the intangible character of digital avatars is easily accepted, within the context of artefact studies we are sceptical of how a lack of existence can be compensated for by aesthetic representation or by virtue of it being open to experimentation and manipulation. Real objects are also open to such interrogation, but with increasing student numbers and pressure on academic staff this comes at a price that society does not seem to want to pay. This is an important point for it allows us to evaluate the true value of digital representations. The experiences provided by the CONTACT project suggest that digital representations are an economically effective means by which to introduce aspects of artefact study to large numbers of students. However, there is a need to emphasize very forcibly that digital rendering cannot and must not be considered a replacement for the tactile encounter that real object handling offers. We are especially concerned with what appears to be a current blind acceptance of the benefits of digitization within the heritage sector under the banner of 'access'. Proponents of digitization such as Renaud seem to envision a perfect contemplation as the ultimate result of digitization, but a more worldly glance at the repertoire of digitized collections would suggest that teaching or research at a more prosaic level is more often what is envisioned by their creators.

There are real questions relating to the successful use of such collections in tertiary education. Recent reports (Green 2006; Harley et al. 2006) indicate that users are more likely to turn to *Google Images* than search specialist online collections for images for use in teaching, while Harley characterizes the philosophy of many online collections as 'supply-driven' and 'build it and they will come', noting 'general lack of knowledge about the level and quality of their use in educational settings' (Harley et al. 2006). Another key concern cited is that available collections do not match real pedagogic needs, and staff always look to mix, reorder and reuse material in teaching rather than to use digital collections themselves in teaching. This is the inevitable result of digitized collections aimed at general or multiple audiences, rather than a wide availability of collections tailored for the curriculum.

Digitized collections are a re-creation in the virtual sphere of existing physical collections, and as such they extend the traditional museum paradigm – look but don't touch – into this new realm. This new level of 'access' has its obvious advantages but does not tackle the key problem that traditional museum displays raise in the pedagogic context: the monitor screen recreates the physical barrier of the glass screen and thus the tactile experience remains as distant in the virtual world as it does in the museum display. The acquiescence of heritage workers to the stampede of digital access projects means not only that limited funds are diverted to support this cause from care of traditional collections but moreover the resources to present real collections to scholars and public alike may be lost. The assumptions that underpin vast digitization projects have rarely been opened to effective critique.

We feel there are significant issues with the pursuit of ever-better digital rendering that need to be addressed by those responsible for initiating digital collections, especially in the domain of teaching and research. Such objectives are ill-conceived and are driven by a flawed understanding of human perception rooted in cognitive science. Cognitive science emerged as an alternative to behaviourism in the 1950s alongside the development of the digital computer (Ingold 2000: 163). At the heart of the cognitive model of perception is the idea that the world 'out there' is represented in the mind as a model, and that such representations are the result of computational processes working on information received by the senses. In this unilinear processual model of perception the validity of the representational model is primarily determined by the 'character' of the stimuli reaching the sense organs. Much like the popular rule of early computing 'nonsense in: nonsense out', our perception of the environment can only be as good as the data input from it. These models and understanding of perception have underpinned the pursuit of digital perfection in the misguided belief that to attain ever higher digital resolution and/or virtual manipulation is to provide ever better primary input stimuli which, in turn, allows for the justified supplanting of the real world with a virtual one. The concentration on imagery is not surprising since cognitive science has tended to privilege vision as visual stimuli are often considered to be 'the richest source of information' (Sekuler and Blake 1990: 21). This elevation of vision amongst the senses has not only created a hierarchy of the senses, it has fetishized the image whilst fragmenting the way in which perception and sensory organs are understood. Despite this, even in cognitive science, the critical relations between different modes of sensing are acknowledged. The relationship between touch and vision is especially critical for our discussion here and applies directly to the teaching of artefacts. It is only by understanding the relationship between touch and vision that we can truly begin to appreciate why an emphasis on digital rendering is misplaced.

The relationship between touch and vision is perhaps most obvious when we ask to look at something and immediately reach out with our hands to touch. Touch offers us much more than simply detecting the presence or absence of something. It provides an understanding of shape, size and weight, and it is through this that we develop understandings of properties such as density; all key properties in the exploration of artefacts. In addition, surface texture is often a critical consideration of which touch can give us a unique understanding. Whilst vision has been called 'the richest source of information' (Sekuler and Blake 1990: 359), touch can be construed as the most reliable of sensory modalities (ibid.). This is especially evident when senses conflict: it is usually touch which is the ultimate arbiter and certainly more trustworthy than sight (ibid.), at least in everyday practical situations. This has long been acknowledged and as early as the eighteenth century the philosopher-cum-cleric George Berkeley, in considering the Molyneux problem, believed that touch provided the basis for calibrating and interpreting visual data (Rock and Victor 1964).

One only needs to observe a student looking at an artefact for the first time to understand the relationship between touch and vision. The hand and eye work in tandem: as the hand explores the eye follows and vice versa. This very dynamic act of exploration relates to the contrasting natures of visual and touch acuity. Whereas foveal vision is most acute when the eye and object are stationary, touch acuity is best when fingers move over an object. Katz (1925) was amongst the first to note this phenomenon and observed that holding the fingers still dulled the ability to sense a surface. Because the visual examination of an object is, to a greater or lesser extent, led by the touch of exploring fingers there is a very real problem for how we translate exploration in the real world, where touch acuity is so important, to the virtual world where the hands are desensitized in favour of the visual.

Caution is needed here not to reduce touch to a simple physical stimulus that guides visual enquiry. We have mentioned already how it is through touch that a sense of size, weight and density is realized. Such understandings rely on more than just the contact of the fingers and hands. Whilst holding an object in the hand a relationship is established between the object and the body: how the object relates to bodily geography of an individual is crucial for developing a sense of size and weight and it is how such memories are assimilated into bodily practices that an understanding of the world emerges (Ingold 2000: 166). This is a point well understood within ecological psychology, and it is on this point that it departs radically from the perceptual model adopted by cognitive scientists. Ecological psychologists begin with the proposition that perceptual activity consists, not in the operation of the mind upon the bodily data of sense, but in the intentional movement of the whole being (mind and body) in its environment. Movement is critical. Whilst cognitive science assumes a static perceiver of external stimuli who needs to reconstruct fragmentary data in terms of representations to gain a 'picture' of the world, the ecological model has a perceiver identifying consistencies underlying the continuous modulations of the sensory array as one moves from place to place. The ecological model (Gibson 1979) then sees perception as accessing the world directly; the perceiver has no need to reconstruct the world in the mind in line with pre-existing schemata. There are significant implications for pedagogic practice and digital resources when perception is understood in this way. Since perception is seen as an active task as opposed to a passive state of being, it can be recognized as being an exploratory process of information pickup which, instead of processing sensations, relies on the continual and harmonious movement, adjustment and reorientation of the full range of sensory organs. Gibson emphasized the importance of actively 'looking, listening, touching and sniffing' as opposed to seeing, hearing, feeling and smelling. So if perception is an active task, then how we perceive depends very much on how we act (Ingold 2000: 166), and it is this which should be the central consideration when we consider the implications of digitizing objects and how we work with them.

To employ digital avatars in a virtual environment and then expect students to explore such objects effectively, no matter how perfect the imagery, is to expect students to perceive and understand the world in a very different way. No matter how visually similar a digital rendering may be, the perception of that object is radically different to its real world counterpart and it is not necessarily possible to translate practices between virtual and real worlds. This is a critical point because if this argument is accepted then the very basis of digital collections for use in teaching can be called into question. Proponents of virtualizing technologies may well suggest that in time there will be technological fixes with haptic gloves and 3D visors which perfectly 'mirror' the experience. Although digitization is constantly becoming more affordable, it is unlikely that full rendering for all sensory aspects will be achievable in the near or medium term. Moreover these future horizons are not directly relevant to the limitations of current digital teaching collections.

There are now numerous online artefact databases available for use in teaching. Most of these are, of course, designed for research: the encounter between the individual and the database is that between a knowledgeable user and a familiar dataset. In light of our critique of current digital collections, it is needful to ask what, if any, is the potential for digital collections in the teaching of artefacts? We have critiqued the digital avatar as an insufficient model for the real experience of engagement with an artefact, but where the student has prior direct experience with that artefact, their perception of the avatar will be crucially mediated by their memory of the real object. Herein lies the real value of digital collections: not as a primary means of engagement with objects, but as a perceptive *aide memoire* to partly maintain the moment of engagement beyond the classroom or museum door. This is a point appreciated by Renaud: '[In] Education ... digital avatars play an irreplaceable and even founding role. They make the work, as idea, available before and after its presence' (Renaud 2002: 16). This is mirrored in the notion of the 'ubiquitous museum' (Sumption 2006). This is a point we are happy to concede, whilst we have issues with much of the aspirational exuberance associated with digitization the proposition that they have value as *ideas* is indeed true. We would go further, however, and say that it is the explicit realization of where value is located in virtual environments that is of prime importance as it is only through this that the potential of virtual environments can be fully realized.

In tackling these inherent problems of designing a virtual learning environment (VLE) employing digital avatars of archaeological artefacts, we realized that there was one aspect of post-excavation analysis that could be effectively modelled in a virtual environment. Whilst the digital renderings of artefacts present issues, the standard segmented process associated with their investigation lends itself to virtualization. This was an important stage in the development of the project as we refocused from the objects involved in post-excavation analysis to the process of post-excavation analysis itself. We felt that the real networks which supported the investigation of real material, coupled with a virtual environment which mapped out a template for how to deal with real material, was an effective and realistic proposition which would directly address the aims of the project and issues identified in the initial survey. The result allows for the student experience to demonstrate templates of process in how to approach real-world encounters with assemblages in post-excavation work, museums, and research projects.

We decided to maintain a commitment to the digitization of artefacts to populate the virtualized process of post-excavation analysis, not so much to provide suitable foci for investigation, but more to enable stages in the process to be easily recognizable between real and virtual environments. In order to model process, the material content of each CONTACT project is divided between a reference collection and an assemblage, such as would be encountered in a real world environment. The VLE itself is divided into sections: the first contains a digitized version of the reference collection (Figure 7.1). This area of the VLE mimics a real reference collection and is most like a traditional online database; it is intended to be used in class in support of real objects. The tutor creates a collection of artefacts representative of the aims of the class and these are presented as 2D and 3D images with associated data. These can be used in class as reference alongside the real objects, and importantly are permanently available and instantly accessible after the class has finished.

The second section of the VLE contains the assemblage. These are the materials chosen by the tutor as the basis of the identification and interpretation exercise to be carried out by the students. Again these are available to the students in class and they must begin their work with the real objects. Faced with an unordered jumble of material, how do you go about making sense of it? This is a question of confidence – a key requirement in approaching a body of material. Evaluations suggest that prior to undertaking this exercise, students don't believe that they are capable of dealing with unknown material, while afterwards their confidence is markedly higher. It is

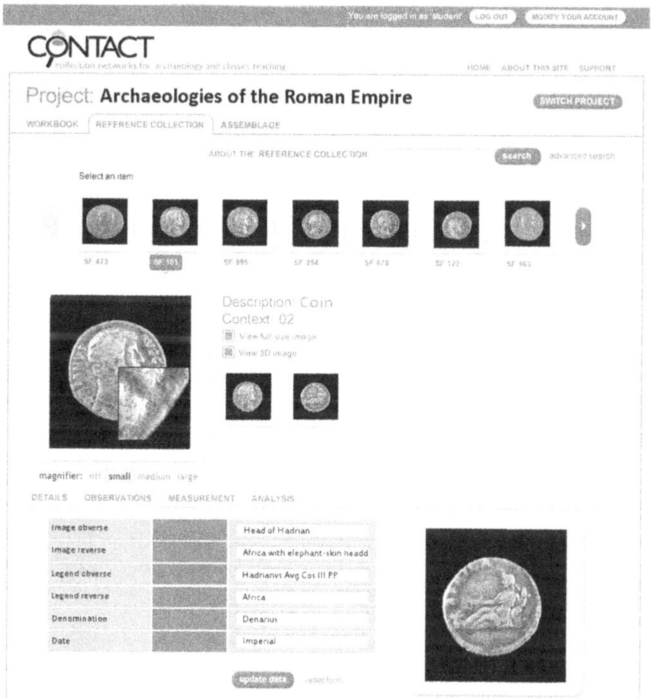

Figure 7.1 Virtualizing coin data in a CONTACT reference collection.

the segmented structuring of the VLE that has proved so effective in communicating the *process* of artefact study. The analogy here – the first stage in the *process* of artefact analysis – is the strewing of a collection of artefacts on a table top – as may well have been done in class. In this section the students can record observations and measurements, according to the tutor's project design – and retain access after class to the digital avatar as augmented by the record they have created.

The third stage of the VLE is the workbook. This provides the students with a structured pathway through the process of developing interrogative strategies and interpreting their data by offering a series of steps in which they carry out tasks and record their results and interpretations. These tasks include quantification, sorting the objects by pre-defined (context, sub-collection) or self-defined groupings (Figure 7.2), carrying out sampling and specialist analysis, and offering high-level interpretations of individual artefacts, contexts and the assemblage as a whole. Tutors can define which of these steps are relevant to their projects from the outset. Evaluations have shown that the structure provided by the workbook steps is popular with students for giving them clear guidance on the nature of the task and how to go about it. It clearly answers our requirement that the VLE should emphasize *process* perhaps more so than that of artefact identification.

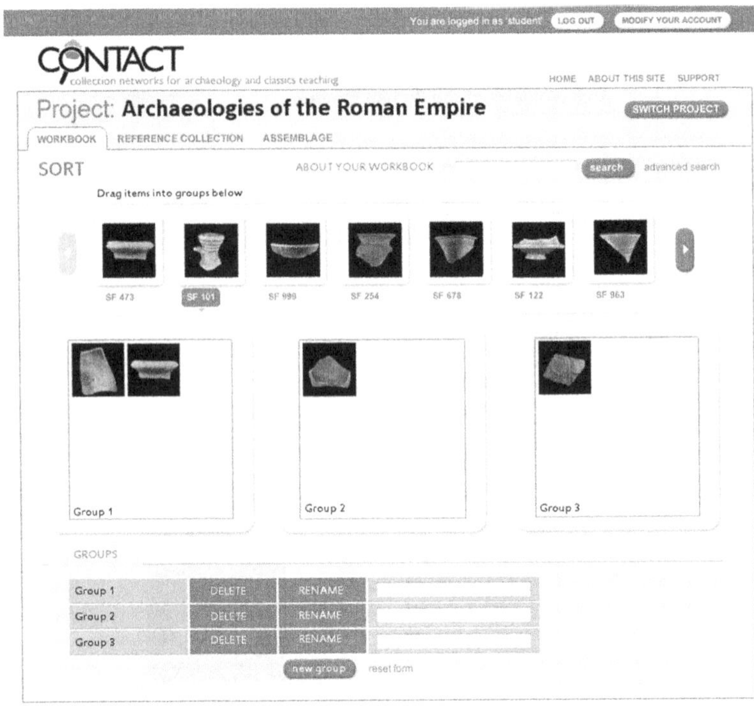

Figure 7.2 Processes of analysis in the CONTACT workbook: sorting objects into groups.

In terms of results and findings, evaluation of the CONTACT pilot projects was strongly positive. Post-session evaluation suggested that the students were confident in their ability to approach, analyse and understand material culture assemblages. In particular, many individuals picked out the *combination* of real and virtual resources as an especially successful element of the sessions – in fact, the CONTACT VLE was regarded as a strongly positive new element in teaching by almost all respondents.

Conclusions

In this chapter we have tried to detail some of the concerns we identified with the digitization of archaeological artefacts in the CONTACT project. This chapter describes the issues surrounding the virtual side of the CONTACT project, deriving as it does from the original Touch workshop on 'new technologies for enhancing object interpretation'. The key word in that title is 'enhancing': the virtual network in CONTACT is not designed to replace real teaching and the real experience of interacting with objects. However, we have demonstrated ways in which new technology can act as an enhancement for teaching and we strongly believe this is in the domain of process and method, and not simply in the digitization of existing artefacts. We have highlighted some theoretical concerns with working with digital artefacts which we are certain will encourage debate as passionate as that between the youths on a Sheffield bus – after all the debate centres on nothing less than how we choose to understand the world we inhabit. As a final word we wish only to repeat the call for a full and frank exploration of the many issues that surround the use of digital media in teaching contexts. With the projected increases in student numbers, already swollen institutions will, we suspect, come under ever increasing pressure to adopt e-fixes to what is otherwise a funding crisis. We only hope that the full implications of such 'solutions' are fully and honestly explored from an intellectual as well as management perspective.

Acknowledgements

The CONTACT project is funded by the Higher Education Funding Council for England through the Fund for the Development of Teaching and Learning.

References

Andrews, K. (2006). CONTACT: Towards a collections network, http://contact.group.shef.ac.uk/contact_survey.pdf, accessed on 8 February 2008.

Chitty, G. (1999), 'Training in Professional Archaeology: a preliminary review. Archaeology Training Forum, http://www.britarch.ac.uk/training/survey.html, accessed 8 February 2008.

Darvill, T., Doonan, R., Pearce, M., Philip, G., Ralston, I. and Sinclair, A. (2007), *Archaeology 2007*, Quality Assurance Agency for Higher Education, http://www.qaa.ac.uk/academicinfrastructure/benchmark/statements/Archaeology.pdf, accessed 8 February 2008.

Gibson, T. (1979), *The Ecological Approach to Visual Perception*, Boston: Houghton Mifflin.

Green, D. (2006), 'Using Digital Images in Teaching and Learning: Perspectives from Liberal Arts Institutions', *Academic Commons* (October 2006), http://www.academiccommons.org/imagereport, accessed 8 February 2008.

Harley, D., Henke, , J., Lawrence, S., Miller, I., Perciali, I. and Nasatir, D. (2006), *Use and Users of Digital Resources: A Focus on Undergraduate Education in the Humanities and Social Sciences*, Center for Studies in Higher Education (CSHE), University of California, Berkeley, http://cshe.berkeley.edu/research/digitalresourcestudy/report/digitalresourcestudy_final_report.pdf, accessed 8 February 2008.

Haselgrove, C., Armit, I., Champion, T., Creighton, J., Gwilt, A., Hill, J. D., Hunter, F. and Woodward, A. (2001), *Understanding the British Iron Age: An Agenda for Action*, Salisbury: Wessex Archaeology.

Ingold T. (2000), *The Perception of the Environment: Essays on Livelihood, Dwelling and Skill,* London: Routledge.

Katz, D. (1925), 'Der Aufbau der tastwelt', *Zeitschrift fur Psychologie* 11.

Osborne, R., Phillippo, S., Powell, J. and Smith, C. (2007), *Classics and Ancient History (Including Byzantine Studies and Modern Greek)*, Quality Assurance Agency for Higher Education, http://www.qaa.ac.uk/academicinfrastructure/benchmark/statements/Classics.pdf, accessed 8 February 2008.

Renaud, A. (2002), 'Memory and the Digital World: A Few Philosophical Pointers for New Memory Practices in the Information Era', *Museum International* 54(3): 8–18.

Rock, I. and Victor, J. (1964), 'Vision and Touch: An Experimentally Created Conflict between the Two Senses', *Science* 143(3606): 594–6.

Sekuler, R. and Blake, R. (1990), *Perceptio*, 2nd edn, London: McGraw Hill.

Sumption, K. (2006), 'In Search of the Ubiquitous Museum: Reflections of Ten Years of Museums and the Web', Museums and the Web Conference 2006, http://www.archimuse.com/mw2006/papers/sumption/sumption.html, accessed 8 February 2008.

–8–

Out of Touch? Digital Technologies, Ethnographic Objects and Sensory Orders

Graeme Were

Over the last decade or so, anthropologists have turned their attention towards the senses as a means to better understand ways of knowing in society. These studies tend to recognize that every society has its own sensory order and that material culture gives expression to this with every object embodying a culturally salient, sensory combination (Classen 1993; Howes 2003). At the same time, ethnographic museums are increasingly recognizing the value of multisensory engagements in learning and are developing special 'hands-on' spaces for accessing their handling collections. Paradoxically, as museums implement such initiatives for enhancing tactile and sensory experience, this rekindling of the sensual relations of material culture appears to be under threat with the emergence of digital technologies such as virtual exhibitions and online catalogues that seem to privilege visual experience at the expense of the multisensory.

This chapter addresses this apparent conflict of interest by investigating three-dimensional digital representations of ethnographic objects housed in a university museum teaching collection. It aims to throw further light on the image politics of replicant technologies in the age of digital technologies to challenge claims that reproductions of original objects are simply shed of their 'aura' (Benjamin 1968), and to show that in fact the multisensory properties of original objects are carried into the virtual world of three-dimensional digital replications. It takes as its focus a pilot project that was set up to augment access to UCL's ethnographic collection for undergraduate students through the development of a virtual museum and exhibition space in a virtual learning environment (VLE). While the project provides insights into how students analyse ethnographic objects scanned in as three-dimensional digital representations and curate their own virtual exhibitions outside the material culture laboratory, it also reveals how digital images possess the same attributes as their material counterparts. By analysing student responses and the virtual exhibitions they created as groups, the project raises fundamental questions about the nature of digital images as it reveals the ways in which three-dimensional images become mediators of the senses under specific conditions. This chapter thus provides further support to question the complex interrelation between the real and the virtual, the multisensory aspects of the immaterial and, in so doing, highlights how further research into the relation between digital technologies, the senses and the virtual museum is urgently needed.

Multisensory Aspects of Ethnographic Objects

Digital technologies facilitate many kinds of connections: connections between museums, visitors and scholars; between different institutions, and between scholars themselves (Hawkey 2004: 3). Indeed, one of the most exciting aspects of digital technologies is the manner in which objects become abstracted from the controlling environment of the museum institution and the exhibition space, free from conservation control and curatorial authority, and physically relocated in cyberspace for people to interpret themselves. Such technologies open new opportunities for bringing together a new kind of knowing that embraces both an analytical understanding of collections through online learning and the sharing of knowledge that augments real-life engagement with material culture. Yet, as visitor numbers to many museum websites soar beyond actual visits to real museums (Hawkey 2004: 10), how well are museums equipped for understanding the potential scholarly benefits of access to their collections in the virtual realm rather than in the real world of the museum space? With the rise of digital content relating to museum collections widely accessible to people all over the world, we urgently need to consider the relation between the real and the virtual in terms of our analytical engagement with objects (material/real and immaterial/digital) and the future implications this may herald for exhibition, interpretation and issues of cultural property. Given issues of access and institutional control within the museum environment (see Bennett 1995), as we enter the age of digital technologies, what will be the future for the real museum and how real will the future museum be? One way to approach this is to understand better the nature of digital images in a climate where digital technologies are increasingly providing capacity to replicate analytical and investigative tools.

Digital technologies pose a particular challenge to our analytical understanding of material culture – both in an anthropological and museological sense – especially so now that heightened importance is placed on the multisensory aspects of the object world. With apparent emphasis on the visual aspects of digital images, with software tools developed to deal with image resolution, detail, zooming, rotating, etc., digital technologies seem to run counter to the sensory turn in anthropology that embraces the tactile, aural, olfactory as well as visual dimensions of objects. These studies have been concerned to critique Western preoccupations with sight and hearing, once indicators of so-called civilized societies, and move towards engaging with the multisensory channels of human sociality with a move away from privileging the visual in social interactions. For example, *The Senses and Society*, a special journal recently launched, seeks to explore the intersection between culture and the senses and shed new light on sensory ways of knowing, and it provides evidence of emerging critiques in the interpretation of material culture.

If digital technologies confine the analytical engagement of objects to their visual properties alone, then this also appears to counter recent museum education initiatives to engage with the multisensory dimensions of objects. Museum handling collections are revolutionizing the way museum visitors learn about collections as educationalists seize on the sensory combinations of objects to offer new kinds of access and engagement, and providing therapeutic benefit to those museum visitors restricted from partaking in the full visual experience of the exhibition space such as the visually impaired (see Candlin 2004; Hetherington 2002). In ethnographic museums today, extremely innovative and exciting educational 'hands-on' resources are being developed with spaces where visitors are given the opportunity to engage with ethnographic objects through a wealth of senses such as touch, sound, smell and the visual in order to learn more about material culture. Many of these innovative learning spaces provide rudimentary documentation and text books, in effect, giving visitors the chance to attain a level of ethnographic knowledge themselves. Two such spaces recently developed are in ethnographic museums in the Horniman Museum, London, and the World Museum, Liverpool.

The Horniman Museum's handling collection was established in 1969 and now numbers over 3,500 objects reflecting the museum's main collections in ethnography, natural history and musical instruments (Martin 1999). As part of their successful Heritage Lottery fund bid in 2001, the museum now has a highly exciting education gallery – the Hands-On Base – which allows visitors to handle objects and engage with them, encouraging full use of their senses. This space is designed to fill the visitor with 'a sense of awe, richness and variety' (Horniman Museum Education Section, unpublished document), with the display spaces in the hands-on area flexible enough to allow the rearrangement of the objects. As such, this innovative learning space is not arranged in classic typological fashion, such as form and function, or geographic zones; rather, visitors are encouraged to engage with objects through material and sensory connections between sets of otherwise disparate objects. For example, a box full of objects is labelled simply as 'red' – and inside one finds a diverse range of objects of varying materials from different parts of the world coloured red. The aim of this space is to provide maximum physical, sensory, intellectual, cultural and emotional access to its handling collection, which, in turn, serves as a mechanism for engaging with the core collection on display in the museum galleries.

Another innovative handling space is the Weston Discovery Centre in the World Museum Liverpool. There, the education service began in 1884, and so one could assume that the handling collection dates back to some time soon after that time. Like the Horniman Museum space, the Discovery Centre encourages multisensory engagement with the objects. While many objects are easily accessible in very robust boxes and drawers, many others features are in glass display cases which are inaccessible to visitors. Similarly to the Horniman hands-on gallery, the

World Museum encourages physical, sensory and intellectual engagement with the collections, as well as an appreciation of the cultural and ethical sensibilities of particular cultures towards their cultural artefacts (Liverpool World Museum, unpublished document).

Yet now in the twenty-first century, with new digital technologies at our fingertips, digital reproductions of museum collections are becoming part and parcel of the visitor experience and exhibition process, particularly with the rise in popularity of online catalogues and databases, virtual tours and exhibitions. In this drive to digitize entire collections and provide increasing access to collections that physically remain in storage, we appear to have been seduced by the technical capacities of these imaging technologies so much so that our understanding of the relation between the real objects and digital reproductions appears extremely limited. What then are people's perceptions of digital imaging technologies and how do these technologies shape the way people engage with objects in collections? Moreover, can we detect any new conventions and practices involved in the analysis and interpretation of ethnographic objects online and what are the power relations that impress on them? Most importantly, what characteristics, if any, of a real object are carried through to its digital representation in the virtual world? The Digital Curation project at UCL developed over 2006–7 sought to investigate these questions.

Digital Curation in the University Museum

The Teaching Collection

The ethnography collection in the Department of Anthropology at UCL consists of around 3,000 artefacts of historical, educational and cultural importance from Africa, Oceania, the Americas and Asia. Most of the collection dates back well before the 1930s and includes several key objects that are instrumental to our understanding of anthropological theory, colonial history and cultural heritage – such as the *kula* shell armlets from the Trobriand Islands, written about by the anthropologist Bronislaw Malinowski in the early 1900s.

One of the great strengths of the ethnography collection is the range of artefacts collected from West Africa dating back to the turn of the twentieth century. The collection includes several wooden masks used for ritual performances (including associated photographic plates and film footage), carvings such as *ikenga* figures used as devices for controlling male power, and a multitude of ornaments forged in metal and designed to display aspects of status and identity. These artefacts not only act as an invaluable resource for researching and learning about African culture, but they also serve as an important reminder of the historic links forged between African communities and members of staff in the Department of Anthropology conducting anthropological field research there.

The ethnography collection also houses two sub-collections of national and international significance. UCL acquired a number of artefacts from the Wellcome Institute for the History of Medicine in the 1950s by the then Head of Department in Anthropology, Daryll Forde, who recognized the importance of university collections as teaching collections. The collection also acquired several outstanding artefacts from the Cuming Museum in the 1960s which includes whale bone carvings from Alaska and shell bracelets from Papua New Guinea.

As a university collection, the ethnography collection was originally conceived of as a teaching collection for anthropology students to learn about world cultures. Today, this is more important than ever. The collection is currently utilized in undergraduate and graduate programmes throughout UCL to teach issues of material culture, cultural heritage, conservation and curatorship. For instance, the collection is currently integrated within an undergraduate core course; conservation undergraduate and graduate practical courses; archaeology undergraduate core courses; as well as postgraduate programmes in material and visual culture. A pioneering new course – ethnographic object analysis – has recently been launched built around the collection and is open to research students across faculties at UCL to learn curatorial skills and techniques through engagement with the collection on a range of issues such as ethics and representation, exhibition and display, collecting and disposal, as well as analytical techniques for materials, technique and style.

Digital Curation Project

Over the course of 2006, UCL's ESCILTA (Executive Sub-Committee for Innovations in Learning, Teaching and Assessment) provided funding for the running of a pilot project, Digital Curation, aimed at developing an online resource tool for supporting teaching and learning about UCL's Ethnographic Collection. A collaborative venture between staff in the Department of Anthropology, Media Resources and Information Systems, project members set out to make three-dimensional digital representations of twelve ethnographic objects in the collection in order to provide a 'digital collection' which could be accessed online at any time by UCL staff and students logging onto the site. One of the stipulations required that the digital representations should provide enough sophistication for details of technical features such as form, technique and materials to be analysed through zooming in and out, rotating and moving the digital renditions using a computer mouse, as well as offering the capacity for users to annotate the images with their own interpretations in a virtual exhibition space which could be viewed by others.

With only a small budget, this project used digital photography in order to create three-dimensional representations of ethnographic objects. Objects were set up on a 'lazy-Susan' dinner service device and photographed at set angles before being 'stitched' together with specialist graphics software. This aspect of the project underlined how, on a limited budget and with the availability of a digital camera and the appropriate software, three-dimensional images of high quality could be generated easily as Quicktime movies. The twelve virtual objects were uploaded into UCL's Moodle (www.moodle.org) VLE as this software offered a number of innovative features that made it particularly suitable for this project. Using Moodle's database module, students would be able to work in small groups picking objects for a virtual exhibition, then analyse and annotate their chosen virtual objects directly on the computer in their halls of residence, UCL cluster rooms, or elsewhere.

The Exercise

The pilot project was introduced to first-year undergraduates taking a mandatory material and visual culture laboratory course. Only registered students on the course could access the virtual museum site. In order to prepare them for the task, students attended an introductory lecture on analytical skills and techniques for ethnographic object analysis. Each student then attended a two-hour laboratory exercise in which they were guided through a series of tasks in small groups and given the opportunity to test their newly acquired skills with UCL's ethnography collection. The main task involved students working in groups of three and engaging with a selection of twenty ethnographic objects with the assistance of academic staff present. For each object, they had to describe the object, what it was made from and discuss the process by which they thought it was made, including the types of techniques employed. They were then challenged to think about the relation between the material and the form of the object by examining how the material constrains the object's form. They were next asked to compare the objects and think about how they would exhibit these objects.

The second task mirrored the first. Students were presented with another twelve ethnographic objects from the collection, all of which had been scanned and uploaded onto the VLE. This time, however, students could only engage with the objects for ten minutes. The idea of restricting their time with the objects was to encourage their use of the analytical tools on the VLE rather than carrying out the analysis wholly in the laboratory. Working in the same groups, the students were directed to examine the objects briefly and choose three objects to include for a virtual exhibition. They were instructed that their chosen objects should form a virtual exhibition based on aspects of material, technique and form. Students could then begin to further their analytical investigation and curate their own exhibition in their groups through the virtual museum accessible via the VLE.

To set up the virtual exhibition, each group was instructed to access the three-dimensional renditions of their real objects via the VLE and upload them into their own virtual exhibition space. Here they could further analyse the objects in terms of their form, material and technique as instructed in the lecture and laboratory session, and then annotate each object with their own interpretations as though they were writing text panels and labels in a museum exhibition. In order to alleviate any problems, students were provided with online forums to allow discussion about their exhibitions, which they utilized.

Results

The virtual exhibitions suggested that students had used the three-dimensional imaging tool in the VLE to examine the ethnographic objects they were assigned to investigate. Many students commented that they found this tool novel in that it allowed them a new way to engage with an object away from the laboratory environment, for example: 'The most useful skill I learned was analysing an object that we had no idea what it was' (Material culture student).

Most students found the instructions clear and entertained the idea that they could go back and see the objects at will, whenever they liked, thus: 'Being able to go back and see the objects at will, whenever I liked' (Material culture student).

In effect, this suggested that they themselves had control over the collection, and in a sense took on the role of the curator. However, some students were more sceptical to the understanding that the exercise itself replicated the analytical process in material culture laboratory, and especially so in terms of the exhibition process. One student stated that more curatorial skills ought to be taught as part of the material culture course and that digital technologies could not replace this.

Of course, the pilot project was initiated in order to augment real-life exploration in the material culture laboratory and was in no way meant to be a substitution for it. Due to constraints on staff time, one purpose of the digital curation project was to enable students to analyse and curate their chosen objects in their own time when access to the laboratory is strictly restricted – as one student commented 'The thing I found most helpful was the ability to analyse an object outside of a laboratory environment' (Material culture student).

But the students' exhibitions and their comments on the work produced nevertheless suggests that they enjoyed the freedom to engage with the objects away from the laboratory and that they benefited from being given the authority to play the role of curator. This observation became particularly apparent from students' comments in the laboratory and in the feedback session afterwards. For instance, students found this exercise quite problematic as they felt that they did not initially possess the curatorial expertise and skills to engage with objects; the general feeling was that the objects needed to be tied down to provenance, cultural group and historical significance. In effect, they felt that this specialist knowledge that they lacked was a hindering factor, elusive to them as non-experts, and rested firmly in the domain of the curator.

However, once they understood that the objective of the task was to concentrate more fully on the materials, techniques and formal aspects of the objects, and draw links between sets of them, then many students felt that they could contribute more fully to the exercise. Students could select objects to analyse, rotate the objects in order to see the material, technique, and formal aspects of the object, then zoom in and out on their chosen object. In many of the virtual exhibitions, and in the student responses, interpretations of the objects suggested that students had taken advantage of the imaging technology in their analytical approach and made connections between sets of objects they were curating. In effect, the curatorial role they had taken on gave them the confidence to pick the objects they liked, move them around to explore in detail such techniques as knotting, carving and binding, and compare these techniques to other objects under their control. Many students felt this process allowed them to learn how to describe ethnographic objects much better and feel more confident in approaching material culture and techniques of formal analysis, as well as working together in a team. The following comments were made by individual students:

'… describing what I saw and analysing it in an anthropological, ethnographic and logical sense …'
'… most helpful was how to analyse an object with little information, and how to interpret the object …'
'… how to research an object and interpret it in relation to other objects …'
'… the specific information I learned from looking into my object in the suggested way …'
'… to describe carefully …'
'… how to write about objects. Although I am still not confident in this …'

One student mentioned that the exercise generated too many ideas, complaining that they had problems keeping the description concise. It is almost as if the analytical software tool revealed too much information about the object under investigation. This could be explained by the process of abstracting the object from the collection and situating it in cyberspace where it is falls under the control of the person sitting in front of the computer screen and investigating it through a mouse interface. Indeed, what seems to be important is how this relation to the virtual object via the mouse interface is a completely novel way of engaging with objects in the material culture laboratory. An implication would be to suggest that we are dealing with a new kind of touch linking hand coordination of the mouse with eye coordination of the cursor on the screen to elements on the surface of the object under analysis. Many students were aware of this, mentioning that the exercise allowed them to think about the relation between the real object with the virtual, and to consider the implications of this through the exercise they were involved in.

One of the criticisms of the project by some material culture students was that the transformation from the real laboratory environment to the virtual immaterial space reinstated a visual approach towards analytical techniques, not allowing one to engage with sensory modalities such as touch, smell and sound. Many anthropologists writing on museum spaces have been critical of the power relations invoked by the 'look, don't touch' ideology of the modern museum (e.g. Classen and Howes 2006). Yet comments from students suggest that three-dimensional digital renditions of real ethnographic objects are not simply confined to visual relations, or, as one student put it, staring at digital images on computer screens. Instead, for some students, looking at digital images allowed them to look for ideas associated with the objects – for instance: 'how to look at objects and look for meaning' (Material culture student).

In this sense, being able to move between objects easily and being in control of the objects allowed connections to be made that could not easily be done in an exhibition space, as this is generally done for you by a curator or exhibition design team. In effect, it invokes the playful nature of early collections where collectors would marvel at their newly acquired curiosities, moving them around and arranging them, comparing them and thinking through them – forming new categories and ideas by making connections between otherwise disparate objects (Stafford 1999).

Student responses to their analytical engagement with three-dimensional digital images through comparison, control and manipulation suggests that they felt comfortable working with the objects as though they were 'real' objects actually in the material culture laboratory. One could suggest that the digital objects were treated in the same analytical way as approaches to objects in the material culture laboratory as this engagement with the object through zooming in, rotating the image, and moving it around replicated the analytical skills and techniques students had been trained to follow in the material culture laboratory. As Armstrong (2006) suggests, what is perceived as 'real' about an image is not its resemblance to something that exists externally to the image, but the physical manner in which one interacts with it – through eye movements, looking closely, turning the head and so forth – so that the relation between image and viewer is not just ocular but also experiential. A similar process appears to take place with the digital images produced for this project, whereby students engage with the objects as though they were handling them, using the computer mouse to turn them around and examine closely a particular technical feature. This phenomenological engagement with the digital image gives a sense of hyper-reality.

Yet, while the three-dimensional rendition on the screen may have allowed students to work with the digital image in the same way as analysing a real object, some felt that the digital image did not allow them to analyse physically the texture, weight and feel of the object that they had briefly handled in the laboratory session. This is obviously one of the major problems of digital technologies though, with the development of haptic technologies for the museum and cultural heritage sector, this may soon be alleviated. However, we should be cautious about possible assumptions that sensory qualities of objects are not carried over when moving from the real to the digital. For instance, the curator and artist Deidre Brown questions the distinction between the real and the virtual in the context of Maori collections (Brown 2007).

In the context of Maori cultural heritage, Brown states that a three-dimensional rendition of a real object is not simply conceived as an immaterial digital image on a screen, but it carries with it the aesthetic and spiritual associations that Maori indigenous people perceive in order to make objects of their cultural heritage operative. This is possible because: 'Maori and national cultural heritage institutions already treat moving and still images of Maori objects, people, and environments as if they embody the same original qualities' (ibid.: 79). In effect, the qualities associated with the real object, when scanned and represented in digital format, are reconfigured in the virtual domain through its inalienability from the original or prototype object which itself is embodied with a particular essence. Such a process thus highlights how two often separate domains of the real and virtual may intersect within specific contexts through a complex set of relations.

While students did not raise any concerns on religious or spiritual grounds about the digital replication of any of the ethnographic objects, they were worried that the multisensorial features of the objects would be stripped through the digital curation exercise. However, on assessing the virtual exhibitions composed by the small groups of students on the material, technical and formal aspects of the design of the objects, most groups made reference to material aspects of touch (weight/texture) and, in some cases, the sense of smell. Interestingly, the interpretation of the objects in the virtual exhibitions suggested that students had used analytical techniques to engage and understand the design of their chosen objects in the same way as they would in the material culture laboratory. This would suggest that students were in fact making connections between the material culture laboratory and the VLE, as well as the real and digital object through the imaging technology – with its level of detail and panoramic scope – and this, in fact, allowed students to visually feel the texture of an object surface, its weight and its material properties, and so on. In effect, the digital image became a conduit for transmitting sensory information and experience.

For example, many students chose a woven hat from the Democratic Republic of Congo in Africa for their exhibition. One group wrote that: 'this hat is composed of a flexible but coarse sand-coloured material which is most likely a type of cotton yarns (*sic*)', while another group was more cautious: 'It would seem to be made from a cotton-like material, but it is difficult to determine this over the Internet. I would want to touch it to be able to determine the texture.' In contrast, another group referred to the material of the hat as 'stiff', while another group said 'it is rather heavy for being made of yarn this thin'. They go on to say: 'The hat must have been treated to make the material stiffer to keep the shape of it'. The three-dimensional rendition of the hat appears to generate further discussions around its materiality. Another group states in their exhibition: 'The material is heavy and strong and therefore it is difficult to fall apart and break'. On a sheet of Fijian barkcloth, the digital image allowed the students to deduce that the material, though flexible, was liable to tear easily. One group felt that the barkcloth resembled a papery-like cloth.

Material qualities of shininess were also remarked on. With reference to the hat: 'the hat utilizes a light colour and light material to deflect the hot sun of the region and to protect the wearer's head from the unbearable heat'. Another group focused on a Polynesian fish hook made from mother-of-pearl: 'It shines from every angle and the pearl like quality of the shell support has an iridescent quality'.

Describing a basket from Madagascar as 'rigid' and 'sturdy', qualities that could arguably relate to touch, one group wrote: 'This tight weaving, using lots of woven loops, not only allows the basket to hold small objects securely, but also means that visually each line in the pattern is slightly fragmented which is interesting.' They go on to comment on the Polynesian fish hook as aesthetically pleasing because 'the natural material glows'. Another group referred to the basket as 'a very delicate object, and almost weightless'. They go on to say: 'The basket smells musty, like it had been in storage for quite a while'.

One group framed their objects around the different textures they exhibit. They describe the Polynesian fish hook as '[it] fits comfortably in the palm of a hand and is generally smooth, yet dense. It is bright and shiny, pleasing to look at. Despite its size, it is dense and its main structure is probably made from the thick, dense shell of a medium-sized sea mollusc, which the darkened whorl section shows up.' Another group, analysing a small carved wooden head from Nigeria, interpreted the surface of the figure as 'skin' which had a rough feel to it. In contrast, another group claimed: 'This carving is made from a soft wood. It fits easily into the palm of the hand and is comfortably weighted; it weighs as much as would be expected of an object of that size.' They go on to then say that the figure has a rough and unrefined finish.

On the famous *kula* shells from the Trobriand Islands, one group of students exhibited them as 'hard and cool to the touch', suggesting therefore at least that the digital image aids one to recall the sensory qualities or that the digital image itself carries with it multisensory qualities even though sight dominates.

So how is it possible that digital images transmit such sensory knowledge? Let us begin by assuming that the digital reproduction contains within itself many levels of social experience of varying levels of concreteness, and, in this respect, it is like the real object in many ways. But the digital image also resembles the real object in the sense that the object represented is known by the students through interaction in the material culture laboratory. The students could thus be familiar with the materials and techniques used in making the object.

But there is one important difference in respect of this knowledge. Imagine if we were asked to analyse an object that we had no previous experience of, and the only way to do this was through a photographic representation. Our knowledge of the object is contingent on our social experience, so that in order to begin to interpret it we would need to share some knowledge of the materials, techniques and the social environment in which such an object would be situated (Schutz 1967). This means that any knowledge of something unknown to us through our own lived experience is open to interpretation. Thus, in the case of the object depicted in the photograph, we would have to infer the types of materials and techniques used in its manufacture through a visual analysis, the incisions, colours and form giving some vital clues as to this. We might even have some knowledge of such techniques and materials from previous experiences with similar sets of objects.

However, in the context of the digital curation project, the replicated objects reside in the world of our lived experience – the material culture laboratory. Through direct sensory experience in the laboratory, students are able to engage with the objects and analyse their form, and the materials and techniques used in their making. Thus, the three-dimensional digital reproductions carry with them aspects of sensory knowledge as a retention of past experience. In recognizing objects from the laboratory exercise, students' knowledge of the objects is already organized under categories that can be referred to in their interpretations.

Yet what would happen if the students only briefly engaged with the objects in the laboratory or indeed, just observed the objects – as many did – without actually handling them? In this context, we find that their primary analytical engagement with the objects was mediated through the tools set up in the VLE and performed by the student using the computer mouse. Any material or sensory understanding of the object could be grasped through the 'hands-on' manipulation of the three-dimensional image on the computer screen via the VLE tool by using the mouse to rotate, drag and zoom into particular features on the object's surface. This physical engagement with the object via the mouse brings about a reconfiguration of the ocular-centric relation normally associated with viewing digital images – a transformation from a direct observation into a direct relationship. In this way, students are both looking and touching: a phenomenological engagement with digital images forming part of their social experience. It is through this new kind of touching that sensory knowledge appears to be mediated.

A Case for Further Research

This study has suggested that the working environment of the material laboratory, a scientific space for the exploration, experimentation and discovery through physical engagement with real ethnographic objects, appears to be reconfigured on the VLE: a warehouse of three-dimensional renditions which can be clicked on, analysed and compared at will, whenever one wishes. Digital heritage technologies thus open up new ways of knowing and experiencing a world replicating in images and, as one student claimed, can lead to three-dimensional renditions carrying too much information. Such a comment underlines the need for further research; the sooner we acknowledge the range of sensory combinations through which we come to know, the better equipped we will become to understand the role of images in society (Stafford 1999). As Susan Stewart states: 'visual perception becomes a mode of touching when comparisons are made and the eye is 'placed upon' or 'falls upon' relations between phenomena' (Stewart 1999: 32). In this case, it is through the coordinated movement of the hand and the eye that one touches the images, bringing into effect a new way of analysing them.

Of course, as a pilot project that aimed to set up an analytical tool to augment object-centred research, as well as create a technical and pedagogical tool through the building of a virtual learning environment, there is much more work to be done in the field of assessing from an anthropological perspective the implications of digital images and the material and sensory qualities attributed to them. The important aspect this project goes to underline is that, as others have argued, there is no mutually exclusive divide between the real and virtual worlds, the material and immaterial; rather, these worlds are intertwined in complex ways. Future research into this emerging area of digital heritage technologies ought to establish further insights into these interlocking worlds as well as the critical issues around the new economy of knowledge of images residing in the digital world.

Acknowledgements

The Digital Curation project was conceived and executed in collaboration with Phil Riding (UCL e-Learning Advisor) and Tony Slade (UCL Media Resources). My thanks are due to them for their challenging ideas, inventive solutions and good humour throughout. Gratitude also extends to Louise Palmer (Horniman Museum) and Lynne Heidi Stumpe (World Museum Liverpool) for their kind help in providing information about education and handling resources in their respective institutions.

References

Armstrong, J. (2006), 'Depiction and the sense of reality', *Contemporary Aesthetics*.4, available online at http://www.contempaesthetics.org/index.html, accessed on 1 February 2008.

Benjamin, W. (1968), 'The work of art in the age of mechanical reproduction', in Hannah Arendt (ed.) and Harry Zohn (trans.), *Illuminations*, New York: Shocken Books, pp. 217–51.

Bennett, T. (1995), *The Birth of the Museum: History, Theory, Politics*, London: Routledge.

Brown, D. (2007), '*Te Ahu Hiko*: Digital heritage and indigenous objects, people, and environments', in F. Cameron and S. Kenderdine (eds), *Theorizing Digital Culture Heritage: A Critical Discourse,* Cambridge, MA: MIT Press, pp. 77–92.

Candlin, F. (2004), 'Don't touch! Hands off! Art, blindness and the conservation of expertise', *Body and Society* 10(1): 71–90.

Classen, C. (1993), *Worlds of Sense: Exploring the Senses in History and Across Cultures*, London: Routledge.

Classen, C. and Howes, D. (2006), 'The museum as sensescape: Western sensibilities and indigenous artifacts', in E. Edwards et al. (eds), *Sensible Objects: Colonialism, Museums and Material Culture*, Oxford: Berg, pp. 199–222.

Hawkey, R. (2004), 'Learning with digital technologies in museums, science centres and galleries', *Futurelab Series*, Report 9, available online at http://www.futurelab.org.uk/resources/publications_reports_articles/literature_reviews/Literature_Review205/, accessed 1 April 2008.

Hetherington, K. (2002), 'The unsightly: Touching the Parthenon frieze', *Theory, Culture and Society* 19(5/6): 187–205.

Howes, D. (2003), *Sensual Relations: Engaging the Senses in Culture and Social Theory*, Ann Arbor: University of Michigan Press.

Martin, D. (1999), 'Education case study: Horniman Museum, London', *Museum Practice* 10: 56–7.

Schutz, A. (1967), *The Phenomenology of the Social World*, Northwestern University studies in Phenomenology and Existential Philosophy, Evanston: Northwestern University Press.

Stafford, B.M. (1999), *Visual Analogy: Consciousness as the Art of Connecting*, Cambridge, MA: MIT Press.

Stewart, S. (1999), 'Prologue: from the Museum of Touch', in M. Kwint, C. Breward and J.J. Aynsley (eds), *Material Memories: Design and Evocation*, Oxford: Berg, pp. 17–36.

–9–

A Versatile Large-scale Multimodal Virtual Reality System for Cultural Heritage Visualization

Celine Loscos

Cultural heritage applications have recently made use of up-to-date advances in haptic interfaces. Specifically, virtual reality (VR) applications were built where the user could visualize museum artefacts in stereo vision and interact with them using two-point contact haptic devices, which have been proved to enhance object recognition and manipulation. In this research we demonstrate that the use of such haptic interfaces improves the overall human experience for cultural heritage VR applications. In particular, differences with standard virtual reality interactions were measured regarding several aspects for two different applications. The first application is the visualization of scanned statues presented as part of the Museum of Pureform. The second application is related to the assembly of a Greek column from scanned pieces. The aspects measured were the precision of the manipulation, the time taken to complete each task, the overall ease of completion and the sense of presence during the experience. Even if the applications were presented as part of prototypes and with research-built haptic devices, it was proven that haptics enhanced most of the measured aspects without any cognitive load, and even helped to improve concentration. It is greatly hoped that these studies will encourage a broader use of such haptic-assisted VR systems in cultural heritage applications.

Introduction

In the last decade applications in cultural heritage have realized the potential of new technologies of databases for accessing information and its dissemination to the public. There is now a general consensus that it is essential to create and maintain a repository of museum artefacts and cultural content. While much effort has been put into information database systems, less has been expended on new display technologies for museums. Besides the cost of multimedia installations, one reason for this is the controversy which surrounds the question of whether it is correct to present a representation of an artefact (i.e. not the original artefact) to the public, and whether subjective descriptions should be provided or simply omitted to allow the visitors to forge their own opinions.

Virtual reality and mixed reality (MR) technologies are now becoming sufficiently developed so that simulations of cultural or architectural sites, or virtual environments (VEs), for design are successful in making users feel truly immersed, especially within CAVE-like displays (Cruz-Neira, et al. 1993) utilizing head-tracking. These systems are usually made using four-wall displays, three of them forming a cube and one serving as a floor. On the four walls are displayed stereo 3D scene content. A user stands in the centre of the cubicle, with a head-tracker to adjust the stereo. Coupled with interactive technologies that allow visitors of the virtual sites to make their own choices or perform actions that trigger responses from the VE, these virtual experiences become richer and more interesting. To date, several successful examples of virtual worlds have been developed worldwide for research and design, training, manufacturing and entertainment. However, such environments suffer either from a lack of realism or a low degree of interactivity, due to technological and methodological constraints (Roussou et al. 2003). On the other hand, haptic interfaces (HIs) have become the primary means of interaction with desktop virtual environments. Several experiments during the past fifteen years have shown how these interfaces can strongly improve user performance during the interaction with virtual objects (Kirkpatrick et al. 1999; Buttolo et al. 1995; Hollerbach and Maekawa 1998). The ability to reflect digitally programmed forces, representing the contact interaction with the object, increments the user performances and their ability to understand what is happening within the virtual environment (Wall et al. 2000; Infed et al. 1999). Finally, early studies have proved that the use of a haptic device for virtual reality applications can enhance the sense of presence (Frisoli et al. 2005; Loscos et al. 2004). Presence is commonly expressed to refer to the sense of 'being there' in the simulated environment, such that the user loses awareness of the real world props that support the virtual illusion (Lombard and Ditton 1997). An alternative view considers presence in terms of 'successfully supported action in the environment' (Flach and Holden 1998); in other words, our ability to perform a task in a virtual environment is intrinsically linked to our sense of presence in that environment.

In this chapter, we present two different cultural applications (Loscos et al. 2004; Christou et al. 2006) that were implemented using specifically adapted multimodal MR systems. The two applications use haptic devices scaled to immersive VR. The haptic devices are systems devised to exert a controlled force on the operator's hand in response to the surface of a virtual object, thus eliciting a sensation of touching the virtual object. Two contact points are used to explore and interact. The choice of two-contact points was justified by a study (Jansson and Monaci 2003, 2004; Jansson et al. 2003). The user interaction and experience of using those systems were evaluated.

The Museum of Pure Form

In the Museum of Pure Form, we explore a novel way of presenting art to visitors of a museum, allowing them to virtually touch artefacts in a virtual museum. In order to realize this, the statues are first digitized with a scanner[1] so that they can be placed in a virtual museum. The virtual museum is then displayed on a 3D stereo screen. The visitor uses a purpose-built two-contact-point haptic device, mounted on an exoskeleton, to explore the shape of a piece of art which the visitor would otherwise be forbidden to touch in a conventional museum.

The Museum of Pure Form (Pureform 2001) is a three-year funded project on the IST program of the European Union. The main concept is to allow visitors to touch virtual representations of artefacts presented in museums. Four museums are partners in this project: Centro Gallego de Arte Contemporanea, Santiago de Compostela in Spain, The National Museum, Stockholm in Sweden, The Petrie Museum, London in the UK, and the Museum of the Opera del Duomo, Pisa in Italy. All allowed artefacts from their museums to be scanned and presented to the public in the Museum of Pure Form. The Museum of Pure Form is intended to work in conjunction with a variety of display technologies and visitors are invited to virtually explore the shape of each sculpture.

As a case study, we tested the Museum of Pure Form in a CAVE-like system, called ReaCToR, into which an exoskeleton-supported haptic interface was integrated. The fully integrated installation is shown in Figure 9.1.

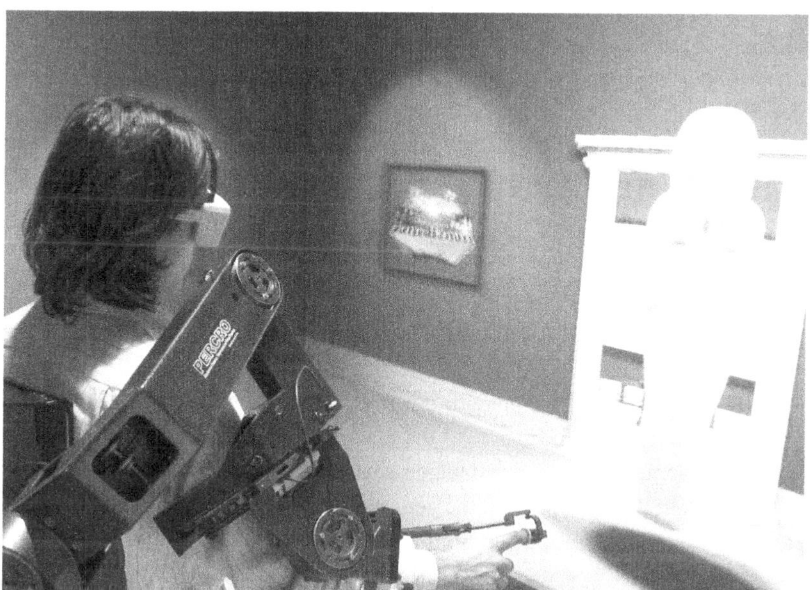

Figure 9.1 User exploring a scanned statue using the Pureform VR system.

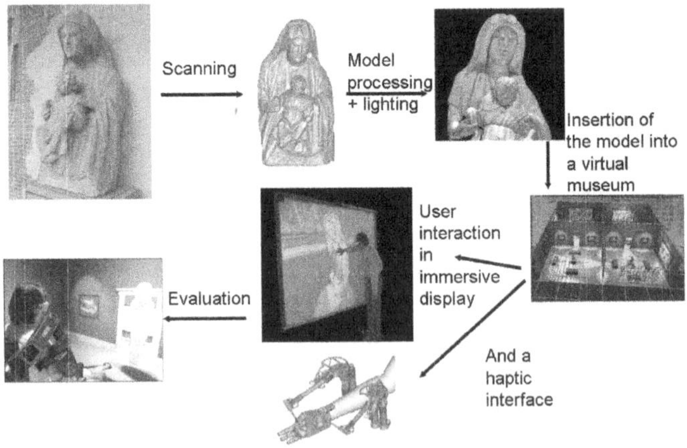

Figure 9.2 The Pureform system overview.

The Museum of Pure Form is a complete system that includes graphics software, a display system, a haptic device and interface, sound, and a virtual museum that contains representations of pieces of art. Its set-up consists of several steps, as illustrated in Figure 9.2.

After geometric representations (models) of the artefacts are acquired through scanning, the raw geometric data is processed to reduce the number of polygons while still maintaining a smooth representation of the artefact. In general, the resulting model is used for both graphic and haptic renderings. The Museum of Pure Form was successfully installed in the ReaCToR. The graphics module runs on a 4-pipe SGI Onyx2 InfiniteReality2. The haptic module and the sound module run on a PC. We observed no user-discernable delay in communication between the haptic and visual components of the system. The haptic device is placed in the centre of the ReaCToR.

Museum Modelling

In order to present the scanned models of statues to a visitor, a virtual museum was created. In this model, a lot of attention was paid to the realism of the rendering, in particular to the lighting. In Figure 9.1 the model of the museum is presented. At the centre of the room, the user can visualize a scanned artefact. The surroundings of the room reflect both the museum and the geographical origin of the artefact on display. Posters illustrating the physical environment of the sculpture's origins are displayed on the walls.

The Haptic Interface

The core systems reproducing the feeling of touch within the Museum of Pure Form are two haptic devices. They are capable of exerting forces synthesized in the digital environment on the user's fingers. The two haptic devices are mounted on an exoskeleton (Bergamasco et al. 2002, 2003). The anthropomorphic exoskeleton HI can be worn on the operator's arm and is particularly suitable for installations in fully immersive VR environments since it avoids optical interference between the projection of the image and the robotic parts of the HI.

Evaluation

Tests were carried out on the usability of the system, the quality of the sense of touch and the overall satisfaction of visiting a museum. A group of computer-literate adults (50 per cent male, 50 per cent female) took part in the experiment. Some of them had visited a CAVE-like system before, but none of them had previous experience with a haptic device. The users wore the exoskeleton and the haptic device, but in one condition the haptic rendering was disabled to allow comparison to when it was enabled. The users were asked to explore the shape of two statues. Each user performed the tasks in a different order. A period of training was allowed for each user to get used to the device. When the haptic feedback was off, the users had to follow the shape with their hand based on visual cues. Each task lasted five minutes. The exploration was accompanied by an audio description corresponding to the statue. After each condition, the users were asked to complete a questionnaire with thirty questions. The questionnaire was designed to elicit the user's experience in terms of usability, sense of touch and sense of presence. The user could answer each question with a number between 1 and 7 inclusive.

Usability

The usability of the haptic device was evaluated through a set of questions on how easy it was to learn to use, the ease of recovery from mistakes, and generally on its ease of use, comfort and overall satisfaction. The mean rating of the usability of the haptic system was 3.2, indicating that some aspects could be improved. In particular the exoskeleton was deliberately set too stiff to ensure safety, making the comfort rating quite low.

Sense of Touch

The users were asked how well they could feel and follow the shapes of the statues, and if they found the haptic feedback realistic overall. Of course, when the haptic feedback was off, the answers from the users were negative. Conversely, with the haptic feedback on, almost all answers were above the midpoint (4). Users found it interesting to be able to touch, virtually, the statues, and found they could feel the shapes well. Despite this they didn't find the sense of touch realistic. This could be expected given that the haptic device provides only two contact points versus the human hand's natural five-finger contact capability with multisensors. Moreover, the interface between user and haptic device was a thimble for each of the two contact points. The comments of the users reflected a feeling that the contact was first on the thimbles and then on the statue. This reduced their sense of reality when exploring the shape of the statue.

Sense of Presence

The questionnaire answers helped in understanding whether the Museum of Pure Form becomes a reality for users, and whether the use of a haptic device was improving this sense. The results are above average for both sets of conditions, with a better score when the haptic feedback is used. This confirmed that the design of the museum, the presentation of the statues and the application itself were appropriate. We can conclude that the museum was well designed, and the representation of the statue appropriate in term of modelling and rendering. We also conducted an additional test that confirmed the results. After the experiment, users were asked four questions concerning the room itself (rather than the statue). They answered with more errors when the haptic feedback had been on, suggesting that they concentrated better on the statue when they could physically touch it.

Conclusion 1

In general we can conclude that users enjoyed the experiment more when they were able to explore the statue haptically. While haptic interaction demonstrably provides a more interesting experience than when using visualization only, the difference in response between those users who experienced haptic feedback and those who did not was not as great as expected. From the user comments, it was deduced that they felt intimidated by the exoskeleton. A CAVE-like system is supposed to allow users to feel at ease. Asking them to wear an exoskeleton is contradictory to the idea of free exploration that a CAVE-like system usually provides. In general, devices should be designed to be less invasive when used in CAVE-like systems.

The CREATE Haptic Interface

In this part we describe the development of a large-scale multimodal simulation to aid learning and appreciation of cultural heritage. This application was conceived within the EU funded project, CREATE. The provision of haptic manipulation of objects in the CREATE project was intended to allow a greater degree of realism and enhanced learning in archaeological demonstrations and for computer-aided architectural design (Drettakis et al. 2007). One innovative aspect of the CREATE project was the integration of a haptic interface with two contact points (Dettori et al. 2003) within a CAVE-like immersive environment in which virtual objects could be projected directly in front of the observer with the ability to grab them and touch them as if they were free-standing. This is a significant advancement over desktop simulations in which the user is more aware that they are manipulating a simulated object projected on a screen. One potential problem to consider with the haptic interfaces is that they could perturb inexperienced users and have the opposite effect to that desired. Therefore, an important requirement for the design of the system was an in-depth study of ease of use.

The particular example (Christou et al. 2006) is that of the Doric temple of Messene, an ancient Greek site that now lies in ruins. The VR simulation allows users to explore and learn about the different architectural members used in ancient Greek temples, specifically their material, positioning and symmetry. This simulation allows users to engage in a process of reconstructing the temple, thus having a 'hands-on' experience of the basic principles involved and architectural components used.

Virtual Modelling of the Archaeological Site

The first step in creating a realistic environment was to capture the real scene. The chosen approach was to acquire the existing site data that would form the basis for construction and reconstruction of the virtual Messene site. The specific monument chosen for our purposes was the temple of Asclepios in the Hellenistic city of Messene (SW Peloponnese). It is a fourth-century BC Doric peripteral temple, relatively large (13 metres wide, 27 metres long and 9 metres high) and simple, built on the remains of an earlier sanctuary of the eighth-to-seventh centuries BC.

Platform

The VR system ran on a SGI (Silicon Graphics) computer, which powered a cubic immersive stereo display, the ReaCTor™ (a CAVE-like environment). A haptic device was installed within the ReaCTor as shown in Figure 9.3, to complement the

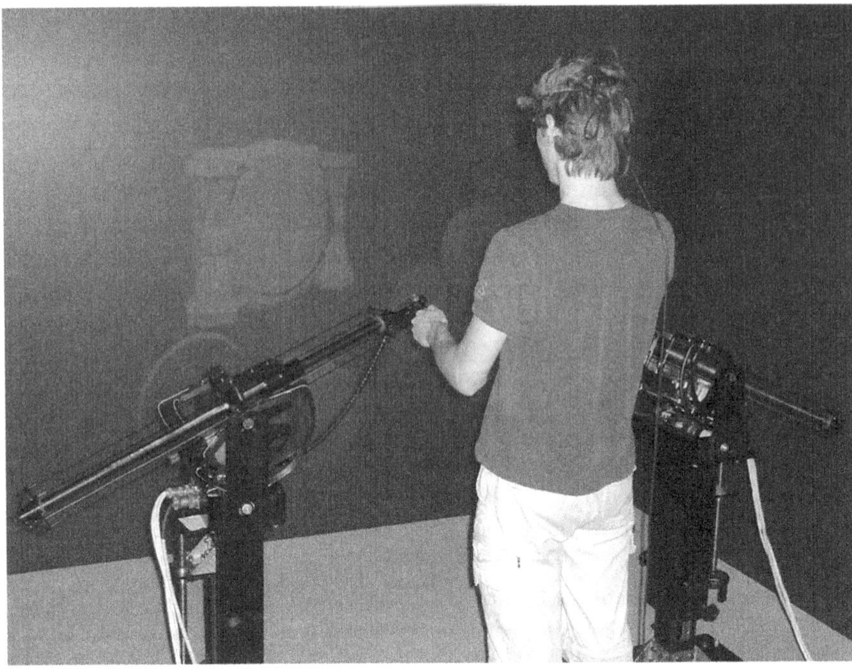

Figure 9.3 Haptic interface shown within a CAVE-like cubic display.

interaction with force feedback and easier manipulation. The haptic installation is further described below.

The interface was intended for use by a wide range of users, including children. It is therefore resizable to adapt to the user's height and is robust enough for use by many different people. It has a large workspace to allow large amplitudes of movement, and the two contact points allow users to touch, grab and manipulate objects in a more natural way. One particularity of the device is that it allows a wide range of exerted forces so as to fit the different tasks, ranging from high frequency texture aspects through to a soft touch to pick forces when grabbing and carrying heavy objects such as columns. Haptic interaction is accomplished using both hands, allowing more natural interaction.

Usability Evaluation

The usability evaluation of the system has taken into account Nielsen's attributes of usability, especially effectiveness, efficiency and satisfaction, and user interface and interaction (Roussou and Drettakis 2005). It consists primarily of the usability of the haptic interface for specific tasks such as handling objects, the environment locomotion principle and usage of the wand for manoeuvring through the environment.

The evaluation of the haptic interface was carried out to gain an understanding of its usage and to improve user experience by using data from the interface itself (Guerraz et al. 2003). The evaluation required inexperienced users to perform a simple task of stacking objects while the time taken was measured, along with other variables.

Evaluations were conducted within the cubical display with projections to three walls and the floor. The display allowed the projection of a scene directly in front of the user, with 3D stereo vision facilitated by the use of CrystalEyes stereo glasses. Head tracking was also used so that when users moved their heads they could observe an appropriate projection of the scene.

The scene consisted of a simulated tabletop, four panels and the three differently coloured blocks that could be manipulated using either the haptic interface or a wand. The wand consisted of a small joystick for navigation (not used in the tests) and several buttons that could be used to make choices. The position of the wand within the display was tracked by the graphics computer and could therefore be used to interact with the virtual objects in the scene.

The virtual table top consisted of three flat coloured panels lying furthest away from the user and marked the start and end locations of the blocks (see Figure 9.4).

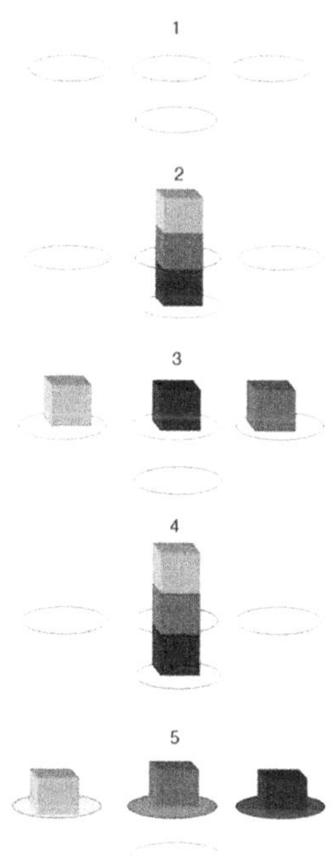

Figure 9.4 Stages in the evaluation test.

Another panel, situated on the table-top directly in front of the user, marked the place where users had to perform the stacking. Users were asked to perform stacking using both the haptic interface and the wand. The haptic interface allowed users to pick up each block and manoeuvre it into place as if it were a real object, whereas with the wand users pressed a button close to each cube in order to select it, then moved the wand in order to position it at the stacking location. The test involved the following:

1. Users initiated the presentation of the target stack by pressing a button on the haptic device.
2. The target stack was shown to the users at the stacking location for 30 seconds.
3. The stack disappeared and the three blocks were presented at the start panels.
4. Users initiated the start of their stacking by pressing a button then recreated the stack as previously seen. When stacking was completed, users signified this by pressing a button.
5. Users repositioned the blocks to the starting positions as indicated by the colour of each panel.

The target stack was created automatically and the colour order of the blocks constituting the stack was randomized. Users repeated the above procedure ten times using both the wand and the haptic interface with appropriate randomization to eliminate biases. The stacking time was recorded as the interval between steps 3 and 4 above. The error in the block placement was also recorded at the end of the trial. Both sets of data were written to data files at the end of each trial.

We recruited nine adult participants from within the laboratory and from external sources. All participants received instructions prior to commencement and also had a chance to perform several practice runs before the actual tests.

The time taken to complete each stack as indicated was recorded in fractions of a second and averaged across all blocks and all trials for each condition (HAPTICS or WAND). The averages for each condition and for each block are shown in Figure 9.5a.

Results show that the HAPTICS condition required on average around ten seconds more than the WAND condition to perform a complete stack. This we attributed to the greater inertia of weighted objects in the HAPTICS condition. Figure 9.5a also shows a significant learning effect with stack times reduced by around 6 seconds in the second session for both the HAPTICS and WAND conditions. Although participants were asked to stack the cubes on top of each other as precisely as possible, there were still errors in placement that could be attributed to ease of use of the interaction tool. We analysed the Euclidean distance between the automatically generated stacking positions of the cubes and the XZ position of each cube as positioned by the user. If users positioned the cubes perfectly on top of each other then the overall error would be zero.

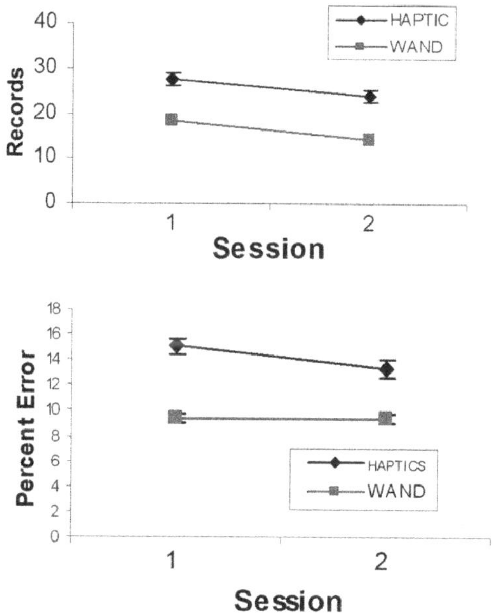

Figure 9.5 (a) Mean time taken (in seconds) to create a stack of three cubes for all nine participants in the study. (b) Results of the usability experiment showing the mean percentage error (with respect to the size of the cubes) for all users.

Figure 9.5b shows the mean percentage error (with respect to the size of the cube) for all users.

The results show that the errors for the WAND were the least for both the first and second session. The errors when using the haptic interface ranged from 15 per cent for the first session to 13 per cent for the second, showing a slight learning effect. Once again, the difference could have resulted from the fact that the cubes in the HAPTIC condition had weight and were therefore slightly more difficult to position precisely than in the WAND condition.

The users had to replicate the stack according to its initial presentation with the correct order of red, green and blue blocks. Errors in stacking could result from not paying attention to the original target stack presentation and also from forgetting the order for placing the blocks. Forgetting is a function of the amount of time between presentation and replication and also of the amount of mental or cognitive overload.

This, in turn, could result from having to do a complex task while the user recalls the original stack order. Errors in stack order were tabulated for both HAPTIC and WAND conditions. We found that in total fifteen ordering errors were made in the HAPTIC condition and twenty-two in the WAND condition. This total is for all nine participants. This shows that the task was relatively easy for all participants and that the use of the haptic device did not appear to increase cognitive load.

These results show small yet significant reduction in stacking time when using the wand. Spatial positioning errors were not significantly different between the wand and haptic interfaces. The differences in stacking times are slight and may be the result of using a physical model in the case of the haptic interface and no physical model for the wand. More importantly, the results indicate no extensive increase in cognitive workload when using the haptic interface, and we believe that this shows it has the potential of becoming a useful component of visualization purposes.

Overall, the haptic interface is shown to be almost as easy to use as the standard wand, with slight increases in stacking time attributed to the greater resistance experienced in manipulating weighted objects with the haptic interface as compared to the weightless objects using the wand.

Final Installation

The preliminary usability study described previously was performed in the formative phase of the design and implementation of the haptic interface. Once the interface was completed and integrated with the cultural heritage application, an additional study was performed with the end users that took place in the Foundation of the Hellenic World's cubic immersive display, during or after normal museum hours (Roussou et al. 2005).

We have presented an immersive multimodal integrated system for evaluation of cultural heritage sites and for architectural design. The system utilizes a CAVE-like display and integrated large-scale haptic interface together with 3D sound to make

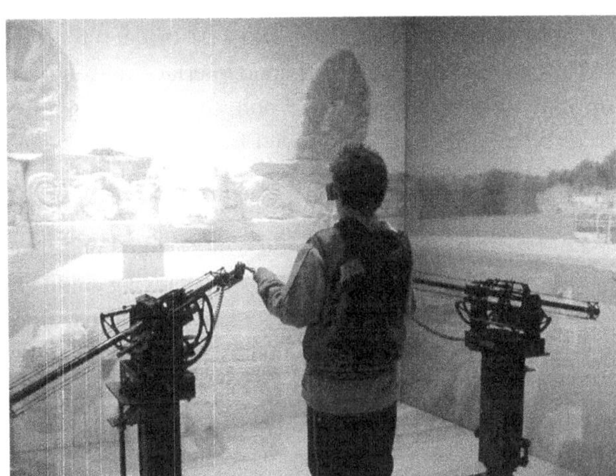

Figure 9.6 The haptic device inside the CAVE-like display of the Foundation of the Hellenic World, used by young museum visitors to manipulate elements of the Messene temple.

the experience as realistic as possible. Our initial project was the reconstruction and visualization of the ancient site of Messene in Greece (Figure 9.6).

The system was used to create an exhibition for the public the aim of which was to teach the process of constructing columns as part of the structure of an ancient Greek temple, but it was also intended for archaeologists to use as a tool for research. The suitability of the haptic interface in this regard was tested under different experimental conditions in order to determine its effectiveness within the given context and its ability to enhance rather than impede the user experience.

Final Conclusions

We have identified issues which need to be explored in the future. Firstly, the exoskeleton and its supports prevent the user from feeling at ease in the environment. The users are constantly reminded that they are in a laboratory, rather than a museum. The design of haptic support structures needs to be investigated in order to produce more user-transparent systems that integrate less obtrusively with CAVE-like systems. Secondly, the development of haptic interfaces is still in its infancy. Using a two-points-of-contact haptic device is a considerable improvement from traditional devices but it still falls far short of the natural haptic sense of a human hand. Combining contact feedback and texture feedback is also an area of research to be investigated. In the Museum of Pure Form only contact feedback was given to the user, whereas the sense of texture would further enhance the experience of users. It would also be interesting to see other applications exploiting haptic devices to augment the experiences afforded by CAVE-like environments. Finally, new interactions metaphors and methodologies should be developed to allow co-located contact between the visual and tactile point as it was proved to increase potentially the user's experience (Swapp et al. 2006).

Acknowledgements

This work was partly funded with PUREFORM (IST-2000–29580) and CREATE (IST-2001–34231), three-year RTD projects funded by the 5th Framework Information Society Technologies (IST) Programme of the European Union. Special thanks to all the individuals from the partner institutions that have contributed to the development of the research projects.

Note

1. 3D Scanners UK; available at http://www.3dscanners.co.uk.

References

Bergamasco, M., Frisoli, A. and Barbagli, F. (2002), 'Haptics technologies and cultural heritage applications', proceedings of IEEE Computer Animation, June 2002, Geneva, Switzerland, pp. 25–32.

Bergamasco, M., Frisoli, A. and Barbagli, F. (2003), 'The Museum of Pure Form', IEEE International Conference of Advanced Robotics (IEEE ICAR03), Special Session on Robotics and Art (July 2003), Coimbra, Portugal.

Buttolo, P., Kung, D. and Hannaford, B. (1995), 'Manipulation in real, virtual and remote environments, IEEE International Conference on System, Man and Cybernetics, vol. 5, pp. 4656–61,

Christou, C., Angus, C., Loscos, C., Dettori, A. and Roussou, M. (2006), 'A versatile large-scale multimodal VR system for cultural heritage visualization', in Proceedings of the ACM Symposium on Virtual Reality Software and Technology, Limassol, Cyprus, 1–3 November 2006, VRST '06, New York, NY: ACM, pp. 133–40.

Cruz-Neira, C., Sandin, D. and Defanti, T. (1993), 'Surround-screen projection-based virtual reality: The design and implementation of the cave', ACM Proceedings of SIGGRAPH '93, pp 135–42.

Dettori, A., Avizzano, C.A., Marcheschi, S., Angerilli, M., Bergamasco, M., Loscos, C. and Guerraz, A. (2003), 'Art touch with CREATE haptic Interface', *International Conference on Advanced Robotics.*

Drettakis, G., Roussou, M., Tsingos, N. and Reche, A. (2007), 'Design and evaluation of a real-world virtual environment for architecture and urban planning', *PRESENCE: Teleoperators and Virtual Environments* 16(3): 318–32, posted online 7 May 2007.

Flach, J.M. and Holden, J.G. (1998), 'The Reality of Experience: Gibson's Way', *PRESENCE: Teleoperators and Virtual Environments* 7(1): 90–5.

Frisoli, A. Jansson, G., Bergamasco, M. and Loscos, C. (2005), 'Evaluation of the Pure-Form haptic displays used for exploration of works of art at museums', in proceedings of *Worldhaptics 05*, Pisa, 18–20 March.

Guerraz, A., Loscos, C. and Ritter-Widenfeld, H. (2003), 'How to use physical parameters coming from the haptic device itself to enhance the evaluation of haptic benefits in user interface', *EuroHaptics* Conference.

Hollerbach, J.M. and Maekawa, H. (1998), 'Haptic display for object grasping and manipulating in virtual environments', in proceedings of IEEE International Conference on Robotics and Automation, *ICRA* 3: 2566–73.

Infed, F. and Lawrende, D.A (1999), 'Combined visual/haptic rendering modes for scientific visualization', in Proceedings of the International Mechanical Eng. Congress and Expo, Nashville, TN, November 1999, *DSC* 67: 93–9.

Jansson G. and Monaci L. (2003), 'Exploring tactile maps with one or two fingers', *Cartographic Journal* 40: 269–71.

Jansson, G. and Monaci, L. (2004), 'Haptic identification of objects with different numbers of fingers', in S. Ballesteros and M.A. Heller (eds), *Touch, Blindness and Neuroscience*, Madrid: UNED Press, pp. 209–19.

Jansson G., Bergamasco M. and Frisoli A. (2003), 'A new option for the visually impaired to experience 3d art at museums: Manual exploration of virtual copies', *Visual Impairment Research*, Geneva: Swets and Zeitlinger, pp. 1–12.

Kirkpatrick, A.E. and Douglas, S.A. (1999), 'Evaluating haptic interfaces in terms of interaction techniques', *4th Phantom Users Group Workshop*.

Lombard, M. and Ditton, T.B. (1997), 'At the heart of it all: The concept of presence', *Journal of Computer-Mediated Communication* 3(2) (online).

Loscos, C., Tecchia, F., Carrozino, M., Frisoli, A., Ritter-Widenfeld, H., Swapp, D. and Bergamasco, M. (2004), 'The Museum of Pure Form: Touching real statues in a virtual museum', *VAST 2004*, 5th International Symposium on Virtual Reality, Archaeology and Cultural Heritage, Brussels, December, pp. 271–9.

Pureform (2001), 'The Museum of Pure Form', IST-2000–29580, 1 September 2001–30 August 2004, available at http://www.pureform.org/.

Roussou, M. and Drettakis, G. (2003), 'Photorealism and non-photorealism in virtual heritage representation', in 4th International Symposium on Virtual Reality, Archaeology and Intelligent Cultural Heritage (VAST 2003) and 1st Eurographics Workshop on Graphics and Cultural Heritage, Brighton, UK: *Eurographics*, pp. 47–56.

Roussou, M. and Drettakis, G. (2005), 'Can VR be useful and usable in real-world contexts? Observations from the application and evaluation of VR in realistic usage conditions', in *HCI International 2005, 11th International Conference on Human Computer Interaction and 1st International Conference on Virtual Reality*, Las Vegas, NV, 2005.

Swapp, D., Pawar, V. and Loscos, C. (2006), 'Interaction with co-located haptic feedback in virtual reality', *Virtual Reality* 6(1): 24–30.

Wall, S.A. and Harwin, W.S. (2000) 'Quantification of the effects of haptic feedback during a motor skill task in a simulated environment', *2nd PHANToM Users Research Symposium*.

–10–

Touch Technologies and Museum Access

Robert Zimmer, Janis Jefferies and *Mandayam Srinivasan*

'Intimate Technologies: Touching Textiles to improve the quality of human–computer haptic (touch and feel) interaction in Material Culture' is an ongoing research project involving engineers, computer scientists and practitioners from the Touch Lab at MIT and Goldsmiths Digital Studios. Our aim is to improve the quality of human-computer touch interaction through the development and deployment of new digital technologies. In this way we will be able bring multisensory experiences to museums and galleries without risking precious and fragile objects. It is one of the principal goals of our research to enable closer approximation to natural touch by providing touch sensation to the user in a way that feels natural, not just through the usual simulation of forces of contact but also through a spatial distribution on the skin that can, for example, mimic the sensation of stroking objects. In particular, we are researching computer-supported access to the feel of textile and other material artefacts, allowing computer users virtually to feel that which is either too delicate or too distant to feel directly. This will lead to a future in which users, researchers, practitioners and curators will be able to access, not just the two-dimensional look of an object, but also its feel. If this works it will transform the museum visitor's experience (both physical and web-based), enhancing relations with museum artefacts and encouraging interactions between museum visitors (both physical and digital)

Introduction

Museums have taken on a dual role: they perform conservation and they enable access. While museums are trusted depositories of important artefacts, which they are charged with preserving in perpetuity, museums are also responsible for providing access to these same artefacts. These two roles can go hand-in-hand; it is impossible to provide access to that which has not been preserved and there is little point in preserving that which is entirely inaccessible. However, the two roles can also be at loggerheads: access can lead to deterioration, the fear of which leads fervent preservers to deny access as much as they can. The uneasiness of the dual role is partially responsible for an evolution, over at least the last 200 years, of fundamental limitations on the kinds of access allowed within museums; museum access is almost entirely visual and solitary. Museums are thus thought of as sterile places in which artefacts are approached as nearly sacred objects of passive veneration, and in which visitors only interact with each other in hushed whispers. Museums were not always so viewed, and with new technologies come new opportunities for access. Indeed, some emerging technologies have enormous potential to enable engagement between people and artefacts and also, using artefacts (or their representations) as a focus, to encourage interactions between people. This potential is only rarely, if ever, realized.

In this chapter, we discuss the kind of place that museums have become, put this in historical perspective, outline the potential that technology brings in principle, describe some emerging technologies, and look towards a future when museums provide real, active engagement, while still fulfilling their mission to conserve.

Museums and Access

The modern museum is all too frequently a site of reverence and silence, filled with people acting and feeling as if they were visiting their dead or moribund relatives. This has been remarked, and regretted, many times – for example, Theodor Adorno commented 'Museums are the family sepulchres of works of art' (Adorno 1967); Merleau-Ponty refers to the museum as a 'meditative necropolis (Merleau-Ponty 1994); Foucault writes of modern museums as places dedicated to 'indefinitely accumulating time', and of expressions of 'the will to enclose in one place all times, all epochs, all forms, all tastes ... constituting a place of all times that is itself outside of time and inaccessible to its ravages' (Foucault 1984). Silent, awed, sight is the dominant mode of engaging with cultural values in our museums or galleries.

But this is a relatively modern view, one that probably dates, as Foucault points out, only to the nineteenth century (ibid.). Eighteenth-century collections, mostly gathered from The Grand Tour, were expressly meant as a means of stimulating enlightened conversation. The cabinets of curiosities that predated these collections, in the early modern period, encouraged multisensory interaction with a great variety of objects. These rooms were arranged with no didactic classification scheme. The collections offered views into a much larger world than they had any other way of experiencing, and the experiences afforded by cabinets of curiosities were far from passive: objects were meant to be handled, smelled, even tasted, as well as seen. To read more about this history, see Zimmer and Jefferies (2007).

It is the goal of our research project to produce and/or deploy digital technologies that will enable us to bring back multisensory experiences to museums and galleries without risking the objects themselves. The general idea is to produce digital models of various aspects of the objects (sight, sound, smell, taste and feel) allowing visitors' experiences of digital simulacra of the objects. For most of this chapter we will concentrate on touch, the focus of this book as well as of our main research. There have been a few experimental exhibitions of touch in museums, including an exhibition that allowed visitors to feel digital images of textured books in Southern California (McLaughlin et al. 2000) and the exhibition called *Touch Me* (2005) in the Victoria and Albert Museum in London in which the sense of touch was the subject of the exhibition itself. However, we are looking towards a more radical and integrated use of digital technologies that will allow more intense interactions with museum artefacts than is presently possible.

To date, the experience of technology in museums and galleries has been considerably less rich than the one we are describing here. While digital technologies have been used effectively to provide context and to enrich the informational content of a museum experience, these benefits can come at some cost. The technology can at times become the principal focus of the visit, distracting visitors from the objects they have come to engage with. Some studies of Personal Digital Assistants (PDAs) in museums, for example, suggest that many people walk through the exhibitions, hardly looking up from their hand-held devices. Moreover, the experience afforded by digital technologies is usually at least as sensually and socially impoverished as it is for unmediated museum and gallery visits.

We are looking to a future in which technology will enable new kinds of engagement. For now, we will concentrate on touch, and in the next section we discuss touch technologies.

Touch Technologies

The general subject of technologies that allow people to interact with each other and with virtual or distant things is sometimes called 'machine haptics' (Srinivasan et al. 2008) or 'haptic interfaces'. The idea of most of the research is parallel to the familiar domain of visual interfaces in which an image of a real or imagined object is rendered graphically as a digital representation that can be manipulated by a computer, or sent from computer to computer, or displayed on screens. In machine haptics, a digital representation of the feel – the three-dimensional contour, hardness, smoothness – of an object is similarly manipulated, transported and broadcast. We will describe some interfaces for inputting or displaying these representations below. Haptic interfaces may some day be as natural to us as visual ones and, if so, they will change our very notions of distance and intimacy. We are only used to touching things that are immediately near us, while we can see and hear things over a considerable distance, even without technological mediation. Moreover, machine haptics, to a much larger extent than visual or aural interaction with computers, has symmetric components defined as:

(a) *haptic displays*: devices that provide simulations of objects suitable for manual exploration and manipulation. These are parallel to that of a computer screen;

(b) *haptic sensors*: devices that respond to information gained from the forces of users' interactions. These are parallel to scanners and cameras that gather visual information about objects.

A full haptic system would take part of both of these aspects and their interaction. A paradigm is described by Srinivasan and Zimmer (2008):

> As an example, imagine a computer-peripheral that had some of the properties of modelling clay so that the model of the shape, textures and hardness of a virtual object would be continually updated using computations based on the forces applied by the user's hands. That recomputed model would then affect what the user feels through the haptic display.

We are not yet able to produce anything like this but we will at some point and it is not difficult to imagine its application in a museum setting; museum visitors would be able to feel the objects from the collection and even get a sense of how they were produced.

Machine haptics can involve both the stimulation and the simulation of the human haptic system through 'virtual environments', in which the human interacts with a computer-based synthetic environment and 'teleoperation', in which the human operates on the real environment at a distance through interface devices. We limit ourselves here to virtual environments to make the discussion easier to follow and because the museum applications are likely to take this form. In the case of virtual environment systems, there are two agents: the human and the computer. Through contact with the haptic device, the human perceives forces to which they respond by providing motor signals that move their hand, which in turn affects the device. The computer detects the movement of the device and responds by sending commands that create the forces that the device applies on the human. Both the human and the computer systems have sensors (nerves, encoders), processors (brain, computer), and actuators (muscles, motors). The machine haptic system mirrors and extends the human haptic system. In the next section we discuss what aspects of the human haptic system need to be modelled and mimicked in order to make future systems that are physically compelling.

What Needs to Be Simulated?

The vast majority of haptic machines concern interfaces with hands.[1] The goal is to create systems that will respond to a user's interactions in ways that are similar to, and ultimately indistinguishable from, real objects when they are explored by a user's hands, either directly or through the use of a tool. The properties of an object that need to be modelled and simulated, therefore, are exactly those properties that people explore with their hands. These include shape, size, and possibly weight, but also surface properties of the material, such as roughness, stiffness and thermal conductivity.

The system must model the user's actions, including (1) moving in free space, (2) pressing an object with a finger pad, and (3) squeezing an object in a pinch grasp.

Haptic Interfaces and Museums

Touching Objects in Museums

The most direct haptic engagements involve touching the artefacts themselves. Many museums of natural history, for example, have objects in which visitors, particularly children, are encouraged to feel. For example, the Field Museum in Chicago and the Natural History Museum in London both have rooms dedicated to physical interactions with objects. This is, not surprisingly, less prevalent in art galleries, though many of these have touch tours particularly for visually impaired visitors. We have been on a touch tour of Tate Modern in London, and other chapters in this book describe experiences of other museum and gallery tours.

Physical Recreations of Objects

One could in theory make physical replicas of real artefacts and invite visitors to touch these. This is a surprisingly rare thing to do, perhaps because of expense and perhaps because of the issues it raises about authenticity and value. Museums seem to be more comfortable making reproductions that are clearly separable from the original, either because they are digital or because they look different.

Raised Pictures

The most prevalent technology (indeed, at the time of writing, the only haptic technology with any substantial presence in museums and galleries) is much less sophisticated than the systems we have been describing so far. The idea is to reproduce pictures so that visual information about the images are encoded as raised bumps, much as in Braille. In this way, people who are visually impaired can sense what the picture looks like. These pictures give a sense of shapes by raising boundaries and encoding information about colour, lighting – information that was entirely visual in the original picture – as raised code.

Tool-based, Force-feedback Haptic Displays

These are the only kind of significant haptic engineering technologies that have (at least experimentally) been used in museum exhibitions. These systems work giving users the sensation of touching a distant or imaginary object with a pen-like tool or stylus. The user holds the stylus and moves it, while the movement is impeded in a way that simulates probing the displayed object with a pen-like object in the real world.

In the PHANToM®, the stylus is attached to the tip of what is essentially a small robot arm. The robot arm has three revolving joints each controlled by an independent motor. The device exerts pressure on the hand through the stylus by sending voltages to the motors, controlling how easily each of the joints can rotate. In this way, force on the user can be exerted in any direction. The computer that is connected to the force-feedback interface takes readings from the three joints roughly once every millisecond. Using this information, the computer can work out the position of the stylus and determine where it maps onto the virtual world, comparing that position with the position of a virtual object being touched.

When the virtual position of the stylus collides with one of the objects in the virtual world, the joints stiffen so that the virtual object seems to resist the user by exerting a force on his or her hands. This attempts to simulate the force that the modelled object would have exerted had the stylus collided with it in the real world, though actual force feedback is limited by the robustness of the device engineering. The user may be prevented from penetrating the virtual object by certain devices, just as the physical object would. By moving the stylus around the space and probing until a force is felt, a user is essentially exploring the space, learning about the contour, hardness and texture of the objects.

Tactile Display Arrays

Tools like the PHANToM simulate the effect of exploring a space or an object with a tool. This is not as tactually satisfying as exploration with hands or fingers, which may be a goal of tactile display arrays. This demands much finer distinctions and more delicate ways of imparting information than the tool-based systems. There are various technologies for trying to achieve this, including shape-changing displays, which convey the local shape of contact by controlling the deformation or forces distributed on the skin, vibrotactile displays, which deliver mechanical energy through an array of vibrating pins placed against the skin, and electrotactile displays, which stimulate the skin through surface electrodes. Vibrotactile displays impart small mechanical signals to the finger tips in an attempt to stimulate the same neural signals from the fingers to the brain as would be sent if the finger had been in contact with an object with particular material properties. This device has been shown to successfully display a wide variety of textures that are perceivable by human subjects, although the present technology is limited. This technology is not yet mature and it does not incorporate spatial resolution that is as fine as that of the human fingertip, where receptors are densely packed. When a user is attending closely to the display, he or she can still feel the individual pins, but this is a technical limitation that we expect to see overcome in the not too distant future. When the technology is mature, we will be able to let people feel materials virtually through peripheral devices. Museums will then be able to let people touch, as well as look at, their collections across the Internet.

Haptic Sensing: Tactile Sensor Array

The two devices already described are both displays providing outputs to the user so that they can feel an object. There are also haptic interactions in the other direction in which the computer gets information from the user's touch. A mouse is a haptic sensing device. The user touches the mouse, pressing buttons and/or moving the mouse's position, and the computer responds to those actions. Compared to the hand, the mouse is a crude device which only responds in a binary way to button pressing. There are much more sophisticated ways of reading the touch of a hand.

One such is an array of pressure sensors embedded in a textile. Pressure exerted on any part of the array alters the electrical properties of the texture by decreasing the distance between the two electrode layers. The system reads these properties and in this way, a profile of how much a user is pressing each part of the material can be computed. If, for example, a material containing this kind of array was used to cover a joystick then the system could gather accurate information about what parts of the joystick were being squeezed at any time, and with what force, allowing for a quite intricate form of computer control. Such devices could be used in a museum setting to give users a sense not just of what an artefact feels like now but also how it may have felt to mould it.

In the above section, we discussed how haptic technologies can enhance users' interactions with artefacts. However, we began the chapter by citing a desire to see museums as places where people also engage with each other. In the next section we describe an exhibition that did that well using sound, rather than touch.

An Example of Technology at Work

The exhibition, which was called *Sounds from the Vaults*, displayed some of the Field Museum's (Chicago, USA) collection of ancient instruments from different cultures. These instruments are being conserved as rare, and sometimes delicate, artefacts and would certainly not be played as a matter of course. Many of these sounds have not been heard for a hundred years. What would be clear to most curators is that it would be important for visitors to be able to hear, as well as see, the instruments. The usual way to do this is to place recordings of trained musicians within the exhibition, possibly accessed through headphones. *Sounds from the Vaults* did have such recordings. But they also had something else; they made digital simulations of some of the instruments. These enabled visitors to make the sounds that the instruments would make – in effect, playing the instruments. The interfaces to the digitized instruments were made easy for an untrained player by replacing the original physical objects with simple electronic interfaces. Visitors engaged physically with aurally faithful simulations of the instruments to make the sounds the instruments could make, producing a technologically enabled, meaningful engagement with the collection.

Furthermore, the curators, Bruce Odland and Sam Auinger, also brought groups of instruments together in the same room and enabled all of the people in the room to hear each other's engagement with the digital instruments. This led to impromptu jam sessions on these simulated ancient instruments. In this exhibition, people used the technology to interact with each other, focused on the sound of the original instruments. We think that this serves as a model for exhibitions. We are looking forward to a future in which museum visitors will, through the mediation of technology, interact with each other using multiple senses in ways that draw attention to important aspects of the artefacts themselves.

Can We Do this with Touch?

In 2002, researchers at the Touch Lab at MIT interacted with colleagues at University College London in what was highly publicized as a trans-Atlantic handshake. What they did was actually a bit different from shaking hands. A computer at each site had a visual and haptic model of the same virtual object. That is, at both sites this object could be seen and felt. What was unique to this setting is that if the object was manipulated at one of the sites the effect could be felt at the other. This was the first real touch-based interaction between people at a great distance and, though we have made it sound straightforward, there were many technical challenges that needed to be met in order to control such problems as latency. It is early in the development of these technologies, but this could eventually lead to systems within which people can interact with each other globally, focused on the look, sound and feel of an object. This would bring back the 'cabinet of curiosity' notion of multimodal interaction and would also realize the eighteenth-century ideal of museum artefacts as catalysts of intelligent conversation – but these would be conversations that can take place across large distances and involve more senses than just hearing.

Conclusions

We are now working towards a future in which visitors and virtual visitors to museum exhibitions will be able to interact with objects and communicate with each other using technologies that allow multimodal interfaces. This is a goal shared by others, including McLaughlin et al. (2000: 19) at the University of Southern California:

> Our mission is to develop seamless, device-independent haptic collaboration such that a museum staff member and a museum-goer or art student at a remote location can jointly examine a vase or bronze figure, note its interesting contours and textures, and consider such questions as Why did the artist make this side rough but that side smooth? or What is this indentation on the bottom for?

In our attempt to reach this goal we are seeking to develop new hybrid, natural haptic displays. Getting these interfaces right requires not only the invention of new technologies but also a cultural study of the way people use technology and materials studies of possible textile surfaces that could be used as the skin of the display. The research trajectory is pleasingly circular as we will explore textiles that can be used in the design of computer peripherals that will, in turn, be used to explore textiles from a distance. To close this loop requires meaningful interaction between engineers, computer scientists, cultural theorists and those working in textiles.

Once we have a natural haptic display, we will be able to touch objects that are either too delicate or too distant to touch directly. The access provided by touch can be extremely intimate and evocative. When we have integrated touch into the digital interaction with collections and catalogues we will have fundamentally transformed museum visitors' relations (both physical and web-based) with museum artefacts and between each other.

Note

1. There is also research on other parts of the human body, including the tongue. The tongue application is both enabled and limited by the tongue's sensitivities. There are many sensors in the tongue but the tongue is quickly damaged by the electrical currents used to stimulate it in these haptic experiments.

References

Adorno, T.W. (1967), 'Valery Proust Museum', in S. and S. Weber (trans) *Prisms 1996*, Cambridge, MA: MIT Press, pp. 175–85.

Foucault, M. (1984), 'Different spaces', in R. Hurley (trans), *Essential Works of Foucault 1954–1884*, Vol. 2, London: Penguin Social Sciences, pp. 175–85.

McLaughlin, M.L., Sukhatme, G., Shahabi, C., Hespanha, J., Ortega, A. and Medioni, G. (2000), 'The Haptic Museum', in proceedings of EVA 2000 Conference on Electric Imaging and Visual Arts, National Museum of Scotland, Edinburgh, 27–29 July 2000.

Merleau-Ponty, M. (1994), 'Indirect language and the voices of silence', in G.A. Johnson (ed.) and R.C. McCleary (trans.), *The Merleau-Ponty Aesthetics Reader*, Evanston, IL: Northwestern University Press, pp. 76–120.

Srinivasan, M.A. and Zimmer, R. (2008), 'Machine Haptics', *Encyclopedia of Neural Science*, Oxford: Elsevier.

Srinivasan, M.A. and Zimmer, R. (in preparation), 'Machine and Human Haptics, in preparation for MIT Press.

Zimmer, R. and Jefferies, J. (2007), 'Accessing material art through technologies of mediation and immediation', *Futures* 39(10): 1178–90.

Part III
Touch and Memory

–11–

A Memory for Touch: The Cognitive Psychology of Tactile Memory

Alberto Gallace and *Charles Spence*

Tactile memory systems help us to process, store and retrieve information regarding those stimuli that impinge upon the body surface, or those objects that we choose to explore haptically (i.e. by means of our hands). Over the last forty years or so, cognitive scientists have investigated a number of the fundamental characteristics of tactile memory using the theoretical framework laid down by earlier studies of visual (and to a lesser extent auditory) memory. This research revealed that people's memory for touch can be sub-divided into a number of different neurocognitive subsystems, some more peripheral and unisensory, lasting for only the briefest of time, others more central and amodal (or multisensory), that help us to remember tactile sensations over much longer periods. This chapter highlights the fact that despite the obvious importance of understanding tactile information processing, surprisingly little research has thus far actually been addressed to the study of tactile memory. What is more, many important topics in this area still remain to be explored. Finally, we show how a better understanding of people's memory for touch may nevertheless provide a number of important suggestions with regard to the potential opportunities (as well as the limitations) that might affect the future use of touch in educational settings, such as in the museum and cultural heritage context.

Introduction

How soft was your first teddy-bear? How smooth/rough was the last dress/suit that you bought? Can you recognize the denomination of the coin in your pocket simply by feeling it? While most people would find it relatively easy to answer such questions, it is important to note that a complex tactile memory system is needed in order to do so. In fact, systems of tactile memory are needed in order to both store and retrieve information about the objects and surfaces that we have touched, and thus facilitate the accurate recognition of objects and materials that we may have previously explored haptically.

Our everyday subjective experience of touching objects (more than simply just looking at them) can provide a particularly effective means of bringing to mind information that might otherwise have been difficult to retrieve (think about the case of resting your feet in hot sand and thus remembering a pleasurable holiday experience). Tactile sensations can also elicit emotional experiences (e.g. see Brang and Ramachandran 2007; Field 2001; Hertenstein and Campos 2001; Hertenstein et al. 2006; McGlone et al. 2007). However, it is only by means of the empirical study of people's memories for tactile information that we can understand why, and by means of what mechanisms, all of these phenomena can occur. It is therefore rather surprising to find that while a large body of empirical research has investigated the mechanisms (and limitations) of visual (and to a lesser extent auditory) memory over the last century, far less research interest has apparently been directed to the topic of tactile memory (see Figure 11.1; Atkinson and Shiffrin 1968; Coltheart 1980; Darwin et al. 1972; Ebbinghaus, 1885; Neisser 1967; Phillips and Baddeley 1971).[1]

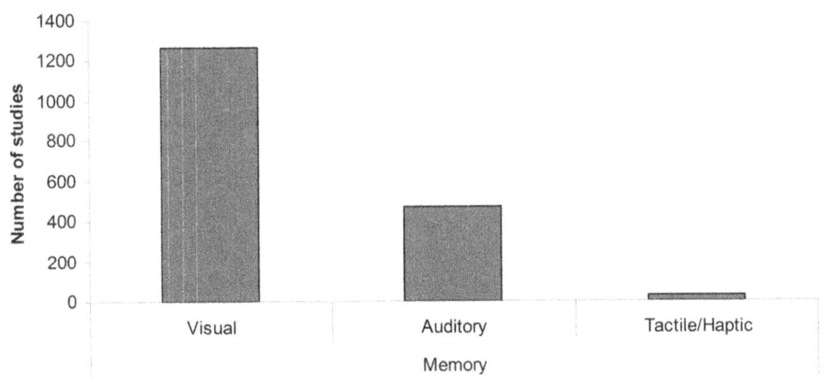

Figure 11.1 Bar graph highlighting the number of studies that have investigated the topic of visual, auditory, and tactile memory (from 1806 to 2007). *Source*: PsycInfo (research date: 12-10-2007). *Note*: only those studies in which these topics are reported in the title are represented in the graph (using the keywords 'visual', 'auditory', 'touch' or 'haptic' and 'memory').

It has been suggested that touch is more 'primitive' than the other senses, such as hearing or vision (Gregory 1967). One of the primary roles of the tactile system is to monitor the biologically relevant features of the environment by means of our direct contact with the outside world and that is perhaps why the tactile system is also the only sensory system to have a direct connection with the motor system (see Tassinari and Campara 1996). Taylor, Lederman and Gibson (1973) have even gone so far as to suggest that something touched is more 'real' than something seen.

The limits that affect the processing of tactile stimuli appear to be different from, and are typically more severe than, those that limit people's ability to process visual and auditory stimuli (see Gallace, Tan and Spence 2006, 2007a, 2007b, in press; see also Gallace and Spence 2008, for a recent review). Therefore, on the basis of these peculiarities (and limitations) of the tactile system, one might legitimately question the extent to which the constraints that cognitive scientists have previously shown to affect the storage and retrieval of visual and auditory information can also be used to help account for the constraints that limit people's memory for tactile sensations.

A related question here concerns whether the specific modality in which information is presented influences how (or even whether) it will be remembered by an individual? Elsewhere, it has been argued that 'the perceptual conditions in which inputs occur, and the links between perception and output systems, are an integral part of the type and amount of information that is available and how that information is processed and remembered' (Millar 1999: 748). In other words, given that touch differs from both vision and audition in so many important regards (see above), we should perhaps also expect there to be a number of important psychological and neurophysiological differences between our memory for touch and our memories for stimuli that have been presented to one of the other sensory modalities (such as, for example, hearing or vision). We therefore believe that tactile memory represents an important, if at present relatively understudied, research topic to be addressed by cognitive scientists.

It is our contention that the study of tactile memory represents a particularly important topic for study at the present time in the context of helping to define the standards that should be laid down when considering the use of tactile information in museum and cultural heritage contexts. Indeed, it is worth noting here that tactile materials (such as sculptures or artefacts that can be explored haptically, or three-dimensional models of architectural features) are now often presented in educational contexts (for both sighted and visually impaired visitors). Indeed, the question of exactly what aspects of the original these models should capture, and how they should most effectively be used, has already been highlighted (see Candlin 2003). Finally, the recent development of new ways of teaching art to people with visual impairments has also necessitated the development of a better understanding of the mechanisms underlying our memory for tactile events and of the extent to which a person's memory for touch changes following sensory loss/degradation in visually impaired individuals.

Different Kinds of Touch

When considering people's memory for touch, an important distinction should be drawn between their memory for the micro-geometric properties of an object (i.e. the smaller characteristics of surface features/textures) and their memory for object qualities (i.e. for the macro-geometric properties of the stimuli; see Table 11.1).

Although these two forms of memory are by no means necessarily unrelated (e.g. in order to recognize an object haptically, both the object's macro- and micro-geometric properties often need to be recognized; see also the section entitled

Table 11.1 Differences between macrogeometric and microgeometric properties of objects together with the tactile systems involved in their processing

	Microgeometric	*Macrogeometric*
Aspect of the stimuli coded	Texture, roughness, stickiness, and spatial density of surfaces	Shape, size, and structure of objects
Stimuli that can be coded	Objects small enough to fall within a single region of skin	Objects that can be grasped by the hands or limbs
Receptors involved	Mechanoreceptors located in the skin	Integration between information from mechanoreceptors and kinaesthetic receptors placed on skin sites that are not somatotopically continuous (e.g. the fingers of both hands)

'Memory for tactile objects'), this review of the literature will concentrate primarily on people's memory for the macro-geometric properties of tactile stimuli (i.e. primarily those relating to an object's shape, size and structure).

Another important distinction to be kept in mind when talking about tactile information processing is the difference between people's memory for objects themselves and their memory for the spatial positions in which those objects/stimuli occurred (were presented). Indeed, it is important to note that touch not only provides information regarding objects that we actively manipulate but also regarding the positions of the stimuli that happen to touch our body surface (think only of your awareness of the location of the fly that lands suddenly on the back of your neck). This literature review will therefore also discuss those studies that have investigated our memory for tactile locations (since, as it happens, this constitutes one of the better studied aspects on people's memory for touch).

In the following, we briefly summarize the extant cognitive psychology and cognitive neuroscience literature regarding people's memory for objects/ information presented via the skin. We will highlight some of the key insights into the neural and cognitive mechanisms underlying the way in which the human brain stores and codes information regarding the objects with which we interact that have been revealed by research on the topic of tactile memory. We believe that the results of this laboratory-based research ought to be kept in mind when considering the future development of policies regarding the handling of objects in more applied settings, such as museums.

Different Forms of Memory: Peripheral (Sensory) Versus Central Systems

Behavioural studies of people's memory for *visual* stimuli have highlighted the presence of a number of different (and distinct) memory systems each with a different and highly specific function. For example, researchers have made important distinctions between long term and short term memories (e.g. Atkinson and Shiffrin 1968), between implicit and explicit memories (e.g. Brooks and Baddeley 1976; Schacter 1987) and between semantic and procedural memories (e.g. Anderson 1987). Moreover, a peripheral/sensory (modality-specific) memory system of high capacity but short duration (known as 'iconic memory' in the visual modality, e.g. Sperling 1960; and 'echoic memory' in the auditory modality, e.g. Cowan 1984; see Coltheart 1980, for a review) has been differentiated from the more central (i.e. in terms of the level at which information is processed in the cognitive system) forms of memory that are affected by any other information processing that a person may happen to be undertaking at the same time (e.g. Gilson and Baddeley 1969; see also Bliss and Hämäläinen 2005; Bonda, Petrides and Evans 1996; Harris, Harris and Diamond 2001; Mahrer and Miles 2002; Ricciardi et al. 2006; see Figure 11.2). It is worth noting here that while sensory memory seems to retain information

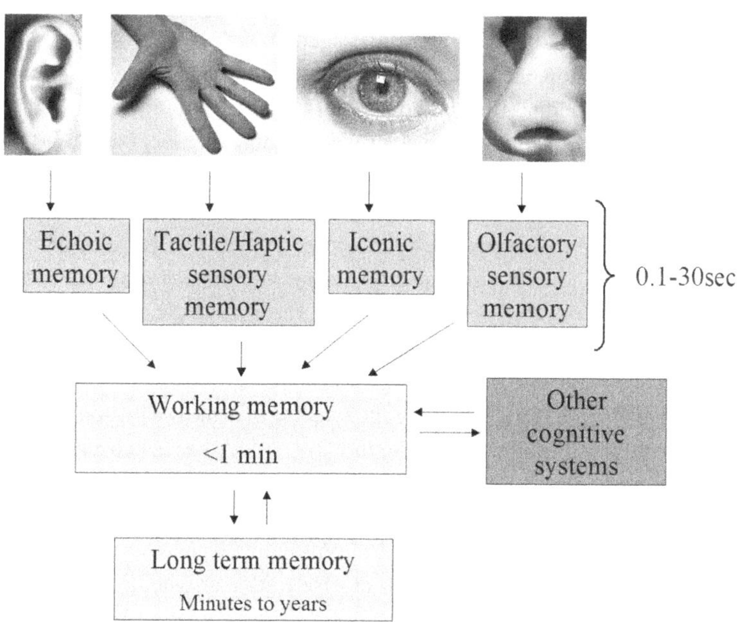

Figure 11.2 Model representing the interactions between unimodal sensory and more central forms of memory (cf. Atkinson and Shiffrin, 1968).

very accurately (although very briefly), and the sensations recalled by this system appear to be phenomenologically very similar to the original stimulation itself (like an 'echo' that retains the quality of the original sound; see Melzack and Eisenberg 1968), the experience of recollecting via other memory systems, by contrast, appears to be very well differentiated from the original perceptual experience.

Memory for the Location of Touch

Although it seems plausible that the same distinctions also apply to people's memory for information presented via the sense of touch, very little empirical research has actually been dedicated to investigating this important (but rather hard to study, at least in comparison to the study of visual and auditory memory) topic. In what may well have been the first study to have investigated tactile memory directly, Bliss et al. (1966) studied the ability of participants to report the locations from which multiple tactile stimuli (presented simultaneously to the fingertips of both hands) were presented. Up to twelve stimuli were delivered to the twenty-four inter-joint segments of a participant's fingers (excluding their thumbs). The participants had to report the digit segments that had been stimulated after each of the stimulus displays had been presented using an alphabetical labelling system.

Bliss et al. (1966) reported that when the participants had to report the stimuli presented to all eight fingers (the whole report condition), the maximum number of stimuli that they could correctly report averaged about 3.6 (out of a maximum of 12 possible correct). However, when a partial report procedure (involving the post-stimulus cuing of individual fingers; equivalent to the partial report procedure used by Sperling 1960, in his seminal studies of visual iconic memory) was used instead, the participants were able to successfully report one additional location (i.e. 4.6 correct).[2] This difference between the results obtained in the partial and whole report procedures was taken to suggest the presence of a "sensory" form of memory for tactile stimuli, one that had a high capacity but short duration (i.e. a form of memory that suffered from very rapid decay). That is, a form of tactile memory equivalent to the iconic memory that had been investigated a few years earlier in the visual modality (e.g. Coltheart 1980; Sperling 1960).

The presence of rapidly decaying storage systems that allow a person to remember the location of tactile stimuli presented to the skin surface has also been investigated more recently by Gallace, Haggard, Tan and Spence (2008). The participants in their study had to report the total number of stimuli presented in a tactile display composed of up to 6 stimuli delivered simultaneously across their bodies (cf. Gallace, Tan and Spence 2006, 2007b), or else to detect whether or not a tactile stimulus had previously been presented in a position signalled by a visual probe given at a variable delay after the tactile stimuli (see Figure 11.3a for the positions on a participant's body where the stimuli could be presented).

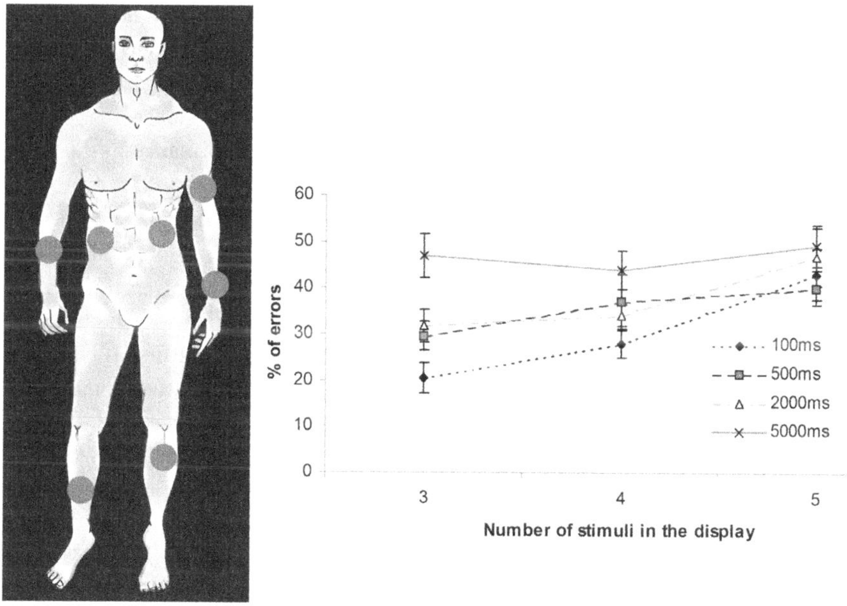

Figure 11.3 (Left) Locations across the body surface stimulated in Gallace et al.'s (2008) recent study of tactile memory. (Right) Results obtained for displays composed of 3–5 vibrotactile stimuli. The graph represents the error rates in the partial report block of trials (where the participants had to report whether or not a particular position of the body was previously stimulated) as a function of the number of factors activated and of the interval between the presentation of the display and the probe for the recall. The error bars represent the standard errors of the means.

The results of the study by Gallace et al. (2008) showed that the participants were only able to report three of the stimuli correctly when they were asked to determine the number of locations that had been stimulated. Interestingly, however, when they only had to report a subset of the tactile locations that had been stimulated, the participants' performance indicated that they were able to access information regarding the location of up to five stimuli (see Figure 11.3). This result shows that short-lasting representations of the location of tactile stimulation can be accessed using partial report procedures, similar to those used previously in studies of visual iconic memory.

Gilson and Baddeley (1969) used a somewhat different procedure in order to investigate people's ability to report the position of a tactile stimulus presented on a single part of the body (i.e. on the participants' forearms) by pointing to it after a delay that ranged from 1 second to 1 minute. The stimuli were presented in one of two conditions: in one condition, the participants had only to remember the position of the tactile stimuli, while in the other condition, they had to memorize the position of the tactile stimuli while at the same time performing a task designed to interfere with any verbal rehearsal strategies that the participants might have been using.[3] This task consisted of counting backwards at their own pace in threes, from an initial three-digit number provided by the experimenter as she touched the participant's arm and throughout the retention interval.

Gilson and Baddeley (1969) reported that participants' performance in this task (measured in terms of the spatial deviation of their pointing responses from the position that had actually been stimulated), was equivalent in the two experimental conditions for retention intervals of up to 10 seconds. However, at longer intervals, performance in the dual-task condition requiring the execution of an interfering task was far worse than that observed in the single task condition. On the basis of these results, Gilson and Baddeley went on to conclude that in the first 10 seconds following the presentation of the stimuli performance reflected the functioning of a more sensory form of tactile memory, whereas the performance observed at the longer retention intervals was related to the operation of a more central memory system instead (see Miles and Borthwick 1996, for a similar result and interpretation).

Evidence in support of the existence of a modality-specific form of tactile short-term memory also comes from a study by Watkins and Watkins (1974). The participants in this study had to try and recall the serial position in which the fingers of their hands had been touched. Participants' performance in this task was disrupted by stroking two of their fingers at the end of a series of the to-be-remembered stimuli (a phenomenon known as the 'suffix effect'). This result was taken to support the existence of a modality-specific form of tactile short term memory, one that is deleteriously affected by the presentation of irrelevant information in the same modality as the to-be-remembered stimuli.

The duration of short-term forms of tactile memory has also been studied as a function of the sensitivity of different body sites to tactile stimulation (e.g. Murray, Ward and Hockley 1975). Researchers have long known that people's sensitivity to detect tactile stimuli varies as a function of the particular skin site being stimulated (it is, for example, higher on the fingertips and on the lips than on the back or on the forearm; e.g. Weinstein 1968). The participants in Murray et al.'s study had to determine whether two tactile stimuli, presented sequentially from one of twelve possible locations on the skin at a variable interstimulus interval, were presented from the same position or not. The authors calculated a 'memory threshold', defined as the distance between the two stimuli that still yielded 'same' responses (the larger this distance the worse the participants' ability to retain information regarding the precise position on their skin surface that had been stimulated). Murray et al.'s results clearly showed that the lower the skin sensitivity measured in terms of participants' two-point discrimination threshold (i.e. the smallest distance between two locations on the body needed to allow two stimuli to be perceived as distinct), the higher was the degree of forgetting the previously stimulated location on the body (i.e. the higher the memory thresholds). This result would appear to suggest that the sensory processing of tactile stimuli and tactile memory are strongly linked (see also Ricciardi et al. 2006, for recent confirmation of this observation using human brain imaging).

Taken together, the results reported in this section provide support for the view that, analogous to the case for vision, there are different memory systems for the remembering of tactile sensations: At least one of these systems is more peripheral and modality-specific (and would appear to be closely linked to the systems used to process tactile sensations in the first place), whereas the other is more central (and can be affected by other forms of processing). Together, these two systems constrain a person's ability to retain information regarding the location from which tactile stimuli have been passively presented on their body surface.

The Memory for Objects that have been Inspected Haptically

Relatively few studies have attempted to address the characteristics and functioning of our memory for objects that have been perceived haptically (as compared to the large number of studies that have addressed people's memory for objects that have been presented visually). Many of the early studies in this area (during the 1960s and 1970s) addressed the processing of an object's geometric properties, particularly regarding an object's shape and size, by the haptic and visual systems (see Walk and Pick 1981, for an extensive review). These studies, in which either two- or three-dimensional nonsense forms were presented to participants, focused primarily on finding any differences that might exist between the ability of participants to recognize stimuli presented to their various senses. The results always showed the haptic system to be very poor, at least when compared to the visual system. Note, though, that in most of this early research, the duration of stimulus presentation was often equated across the various different modalities, thus resulting in a major disadvantage for the tactile modality where long-lasting serial (rather than parallel) scanning procedures (of stimulus exploration) are typically required.

Many researchers have investigated the ability of both sighted and visually impaired individuals to recognize common objects when they are presented in the form of raised line drawings (e.g. Heller et al. 1996; Lederman et al. 1990; Magee and Kennedy 1980). The results of such studies have generally shown that the accuracy and latency of participants' responses in this kind of task vary significantly between individuals (both within and between the different experimental groups) and as a function of the specific conditions of stimulus presentation that have been used. It is important to note here that changing certain parameters of the tactile stimuli, such as their size, can have a dramatic effect on the rate at which they can be explored haptically and, indirectly, on the specific memory demands presented by the task (i.e. more time is required to integrate the different aspects of an object into a complete representation as its size increases, e.g. Berla 1982). Note also that a difference in the orientation in which the stimuli are presented for encoding and subsequent recognition has also been shown to adversely affect people's performance in unimodal tactile and visual object recognition tasks (with visual recognition being deleteriously affected by rotations around the *x*- and *y*-axes, while haptic recognition was affected equally by rotation about all three of the major axes, see Lacey, Peters and Sathian 2007).

Despite the relatively poor performance of people in haptic recognition tasks when using nonsense shapes, their performance reached levels of accuracy that were comparable to those obtained using visual stimuli in those studies where more ecologically valid stimuli, such as three-dimensional objects, or even real faces, have been used (e.g. Kilgour and Lederman 2002; Kiphart, Auday, and Cross 1988; Klatzky et al. 1985). For example, the blindfolded participants in a study by Klatzky et al. had to try and identify 100 randomly presented three-dimensional objects haptically (comprising stimuli such as a book, carrot, egg, hammer, button and candle etc). The performance of the participants in this task was surprisingly good, at around 96 per cent correct.

On the basis of their results, Klatzky et al. (1985) went on to conclude that two-dimensional raised-line drawings may not necessarily provide the best choice of stimulus to use when presenting tactile information displays (for example, in those situations where information about objects or structures needs to be presented to visually impaired individuals, such as in a museum, art gallery, or cultural heritage context). Instead, they argued that effective tactile displays should make use of three-dimensional stimuli that retain certain critical properties of the stimuli (such as their texture). Indeed, Krantz (1972) has proposed that five qualities of tactile stimuli are particularly important for successful haptic object identification: The roughness, size, temperature and sharpness of the object, as well as the amount of exertion needed to explore it. None of these qualities can easily be represented correctly in a two-dimensional display. These observations, emerging from laboratory-based empirical research, should be considered of great importance for those conditions in which tactile materials (objects/artefacts) need to be presented in a museum context and, in particular, for the design of aids for people with visual-impairment.

Interactions between Visual and Tactile Memory: Shared Representations?

A new wave of research interest in the cognitive sciences has recently started to investigate the ways in which different sensory modalities interact when people represent (and respond) to the objects and events that surround them (see Spence and Driver 2004, for a recent review). Unsurprisingly, the way in which people store and retrieve information has not been immune from this trend. Given that the majority of events are 'multisensory' in nature, the investigation of our memory abilities (and constraints) under conditions of multisensory stimulus presentation clearly constitutes a more ecologically valid approach than focusing on each single sensory modality individually.

The participants in a study by Easton, Srinivas and Greene (1997) had to read twenty words, ten of which were presented visually while the other ten were presented haptically. After performing a distractor task for 5 minutes, the participants were presented with 30 three-letter word stems (including the twenty previously presented letters and ten word stems that had not been presented before). The participants had to complete the three-letter word stems, after either viewing or feeling them, with the first word longer than three letters that came to mind. When the letters were presented serially, the 'cross-modal' transfer (priming) of memory was found to be comparable to within-modal transfer. That is, the participants more often used the previously presented words to complete the three-letter word stems when the previously presented letter stems were used, no matter which sensory modality they had been presented in originally. Note that although this result might be taken to suggest the presence of shared representations for different sensory modalities, the nature of these representations is still a matter of some debate. In particular, it might be thought that after the stimuli are first processed by each sensory system individually, the information that is actually stored in memory is semantic in nature rather than being geometric, visual, or even more abstract in form (see Easton et al. 1997, on this point). Different results might therefore be expected if non-verbal stimuli were to have been used instead (such as unfamiliar two- or three-dimensional objects).

Lacey and Campbell (2006) described a series of experiments in which they tried to investigate the nature of the representations that are retained after the participants in their study had been presented with materials either visually or haptically. An interference technique was used in order to determine whether people's memories for visual and tactile stimuli rely on amodal and/or verbal representations. The participants had to look at (visual encoding) or feel (haptic encoding) a series of familiar (everyday household or personal use items) and unfamiliar objects and try to remember them for a subsequent recognition test. They were explicitly told that visual encoding could be followed by haptic recognition or vice versa. The participants were also informed that during the encoding stage of the experiment, different distractors could be presented. Specifically, during the visual-interference condition, the participants also had to look at a visual display in which irrelevant information (dynamic visual noise) was presented. In the verbal interference condition, the participants were instructed to listen to a prerecorded text whose content was irrelevant to the task at hand. Finally, in the haptic-interference condition, the participants had to keep an irrelevant object moving in their non-dominant hand whilst either looking at the relevant objects or feeling them with their dominant hand.

The results showed that the ability of participants to recognize unfamiliar objects was disrupted by both visual and verbal interference, whilst their recall of familiar objects was unaffected by either type of interference. The authors took this result to show that the representation of unfamiliar objects may be verbal in nature. It is worth noting at this point that Lacey and Campbell's results are consistent with the model of object representation put forward a number of years ago by Johnson et al. (1989) that argued for the existence of a serial/cascade process in which haptic inputs activate tactile representations, which, in turn, activate visual representations. If the object is familiar then the activation spreads to the verbal code, triggering the name of the object.

Some researchers have recently started to look at people's ability to recognize complex scenes using either their eyes or hands (e.g. Newell et al. 2005). In one series of experiments, the participants first had to learn a scene composed of up to fifteen objects arranged in random positions. After this learning phase, the experimenter then swapped the position of two of the objects and the participant had to try and identify which objects had been switched. The presentation of the displays containing the swapped objects could either occur in the same sensory modality as the stimuli had originally been presented in, or in the other sensory modality. Note that when the initial presentation of the scene occurred visually, the participants were only given 10 seconds in which to explore the objects, whereas they were given 60 seconds to explore the display haptically (thus compensating for the longer exploration time required by the serial scanning that takes place in the haptic modality as compared to the parallel intake of information that typically occurs in vision). The results showed that performance in the recognition task deteriorated when there was a change in both the sensory modality and the orientation of the scene between the learning and test phases of the experiment (as compared to the condition in which the modality in the learning and test phases of the experiment was the same).

Newell et al. (2005) interpreted this difference in their participants' performance between the crossmodal and intramodal conditions in terms of possible differences in the way in which spatial layout is represented visually versus haptically. More specifically, the authors argued that the previously reported systematic distortions in the haptic representation of space (i.e. certain studies have reported that space becomes more compressed as one moves away from the body midline axis, for example, Kappers 1999, 2007; Kappers and Koenderink 1999) might have affected the performance of their participants in the crossmodal recognition task. This observation might be taken to suggest that different representations are sometimes used to store information regarding haptic as compared to visual scenes.

The results of the studies reviewed in this section, highlighting the interaction between people's memories for stimuli presented in different sensory modalities, support at least three important conclusions: (a) multisensory experiences might lead to 'richer' memories than unisensory experiences (see Lehmann and Murray 2005; Murray et al. 2004); (b) the presentation of tactile information may help people to retrieve information regarding the multisensory qualities of the objects and stimuli that they have interacted with; (c) there are differences between the way in which we code (and store) our memories for objects that we have seen visually versus those that have been experienced solely by means of the sense of touch (i.e. haptically).

Tactile Memory in the Blind

One might wonder whether there are any important differences (in capacity, duration, and/or vividness) between people's memory for tactile stimuli observed in the sighted versus in visually impaired individuals. Indeed, it is worth noting that the people who are visually impaired are generally more experienced in recognizing stimuli using touch/haptics as compared to the sighted. What is more, the research that has been conducted to date has shown that visual deprivation at an early stage of development can sometimes lead to the reorganization of the visual cortex (the area of the brain that is normally responsible for the processing of visual stimuli). In particular, it has been shown that the visual cortex is frequently activated when blind participants perform tactile tasks such as Braille reading (e.g. Sadato et al. 1998; see also Buchel et al. 1998).[4] These effects appear to be related to the fact that the visual cortex, which is normally responsive to visual stimulation, can also respond to inputs from other sensory modalities, such as touch (e.g. Pascual-Leone et al. 1995). Finally, it has recently been reported that many blind people exhibit a superior memory for verbal materials (correlated with the activation – absent in sighted participants – of their occipital cortex) as compared to sighted individuals (e.g. Amedi et al. 2003). Following on from these observations, one might expect there to be differences between the abilities of sighted versus visually impaired individuals to remember tactile stimuli.

Millar (1975) investigated the tactile memory abilities of the blind in a study in which congenitally totally blind children at different stages of learning Braille had to try and recall lists of Braille patterns that were presented sequentially in the exact order in which they had been presented. After the presentation of the lists, the children had to point to the serial position that a given test pattern had occupied in the target memory series (a task testing the participants' memory 'span'). Critically, the lists of to-be-remembered items were: (a) composed of stimuli that were tactually similar; (b) composed of stimuli that had similar sounding names; or (c) composed of stimuli that were dissimilar on both of these dimensions. The results showed that the blind children's recall was impaired by both tactual and verbal similarity. However, phonological (verbal) similarity was found to have a more detrimental effect on performance in those participants who showed faster naming latencies for the Braille items that had been presented in a pre-test part of the experiment. By contrast, the performance of those participants who were slower at naming in the pre-test, were more impaired by tactual similarity. This result shows that after the tactile materials have been associated with specific phonological codes, they were stored in this verbal/amodal form by the visually impaired individuals in this study.

Possible differences between the tactile memory of the sighted and visually impaired were also investigated more recently in a study reported by Arnold and Helron (2002). They asked two groups of participants (ten deaf-blind and ten sighted-hearing participants) to complete four different memory tests presented consecutively in four different experiments. The participants had to identify complex objects (Experiment 1; or 'dominos', Experiment 2) that had previously been explored haptically for two minutes when presented among a group of distractors (in a recognition memory test). In a third experiment, the participants had to try and remember the position in which coins had been placed in a 2 × 2 or 3 × 3 matrix (spatial recall memory test). In a final experiment, the participants had to play a tactile textured version of the game 'matching pairs' (requiring the recall of the position of those pairs of cards having the same texture within a 4 × 3 matrix). The sighted-hearing participants were all blindfolded before the memory tests began.

The results of the four experiments showed that the deaf-blind participants took less time than the sighted-hearing participants to feel and remember the target stimuli. The mean time taken to complete the recall tasks was also significantly shorter (and no less accurate) for the deaf-blind than for the sighted-hearing participants. Finally, the deaf-blind participants also took less time and needed fewer attempts in order to complete the matching pairs game. On the basis of these results, Arnold and Helron (2002) went on to conclude that the deaf-blind participants' encoding of tactile information is more efficient than that of the sighted-hearing participants and that it is probable that their short-term storage and retrieval were normal. This view appears to be consistent with Rönnberg's (1995) claim regarding the compensation of perceptual abilities in deaf and/or blind individuals as a consequence of their lack of visual input:

> Overall, sensory impairment and concomitant perceptual compensation in intact modalities is a cherished myth, rather than a real, robust finding, both for the blind and the deaf. But improvement – in terms of sensory substitution or compensation via impairment-unrelated cognitive functions – represents definite realities. (Rönnberg 1995: 268)

The research that has been conducted to date with blind individuals has also addressed the question of whether or not the presentation of tactile information may improve their memory for acoustically presented materials as well. Specifically, Pring and Rusted (1985) investigated the extent to which two groups of blind participants (twelve congenitally blind and twelve individuals who had become blind two or more years after birth) were able to use picture information presented tactually in order to improve their memory for the content of spoken texts. The results demonstrated that both groups of blind participants showed improved recall for the information represented in both the text and the picture. Moreover, the presentation of the tactual pictures also resulted in the increased recall of pictorial information in the congenitally blind participants. However, it should be noted that this occurred at the expense of the recall of non-pictorial information that was only represented in the text. These data therefore suggest that visually impaired individuals' memory for textual information can be significantly improved by means of the presentation of tactile pictorial information. This observation should therefore be borne in mind when considering the development and use of tactile aids for visually impaired individuals in a museum or cultural heritage context.

Conclusions

The results of the research reviewed here highlight some of the most important characteristics and limitations of people's memory for tactile information. Psychologists have generally tended to use the extant models of memory that have been developed for the visual and auditory modalities in order to show that different memory systems are responsible for different aspects of the storage of tactile information (see Figure 11.4 for a representation of the components and characteristics of a possible tactile memory system that have emerged from the literature discussed here).

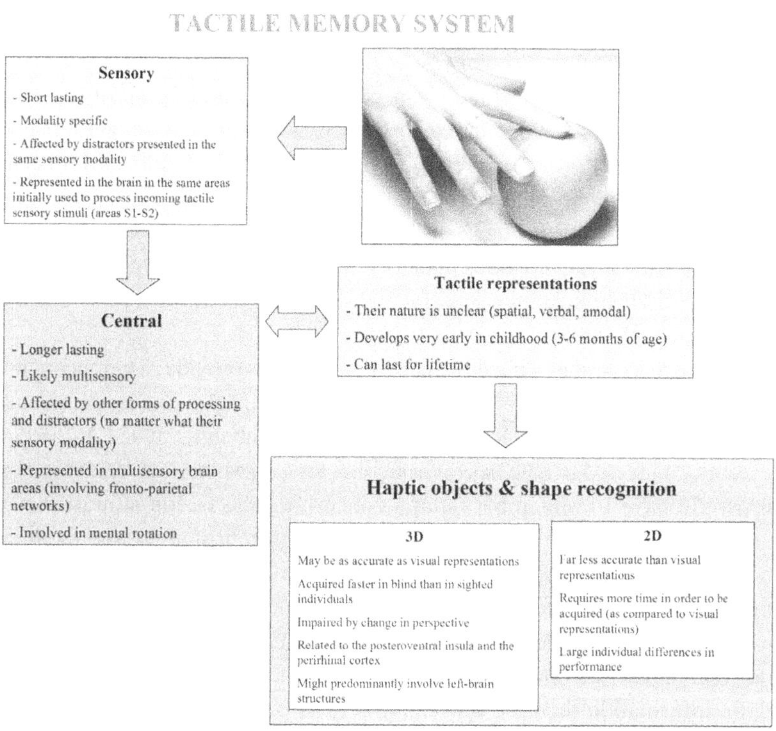

Figure 11.4 Model representing a possible cognitive system for the retention of tactile information.

At first glance, these models would also seem to be compatible with what is known about the organization of people's memory for tactile stimuli (although more research is certainly needed in order to further substantiate this claim). In particular, the research has shown that short-term tactile memory systems appear to be involved in the brief storage of information regarding the position and sensory qualities of stimuli presented across the body surface. Interestingly, behavioural and neuroimaging studies have shown that these short-term representations appear to rely upon the very same areas of the brain that are involved in the original *processing* of tactile sensory information (e.g. Ricciardi et al. 2006). That is, it does not appear as if there are encapsulated and/or separate systems for the short-term storage of tactile stimuli (e.g. Harris et al. 2001; Ricciardi et al. 2006; see also Murray et al. 1975).

Researchers have also highlighted the presence of longer-term representations of tactile information, representations that are generally used to recognize and remember the qualities and spatial position of the objects/stimuli that we manipulate (e.g. the softness of your teddy-bear, the size and feeling of a coin in your pocket, the positions and shapes of your house keys in a dark room or in your handbag, the feeling of a swollen lymph gland by a physician; cf. Esteves, Geach and Spence, submitted). It is, however, important to note that the nature of these representations is still being debated by scientists. Indeed, while certain empirical data appear to suggest that these representations are, at least in part, verbal (and shared between different sensory modalities), other research apparently supports the view that they are actually 'spatial' in nature (and not necessarily the same for all of the senses). However, a large part of the literature reported here would appear to suggest that interactions between the information presented via different sensory modalities occurs at a number of different stages of processing leading to the storage of information in the brain. The suggestion has also been made that when memory needs to be used actively (i.e. during working memory tasks), common representations (with common neural substrates) may subserve the storage of tactile as well visual stimuli (e.g. Ricciardi et al. 2006).

We believe that the study of tactile memory in humans provides a number of potentially important suggestions for the use of tactile information in a museum/cultural heritage context (Field 1975; Koran et al. 1984; Oppenheimer 1972). In particular, the research conducted to date has revealed that longer exploration times are needed when people have to remember details concerning the tactile stimuli that they have manipulated. This is due to the fact that a time-consuming serial exploration procedure (as compared to fast parallel visual exploration) is required for the tactile information to be coded correctly (e.g. Newell et al. 2005). Researchers have also shown that the presentation of three-dimensional stimuli/objects (especially when they are real rather than nonsense objects) leads to the better recognition and storage of tactile information than the presentation of two-dimensional stimuli (such as raised lines drawing; for example, see Klatzky et al. 1985). This observation also seems to be important when tactile information has to be provided to visually impaired individuals for educational purposes. With respect to this topic, it is also worth noting that the use of tactile pictorial stimulation has been reported to improve visually impaired individuals' memory for auditorily presented materials (such as text). This result suggests that tactile aids should be provided together with audio guides wherever possible to potentially enhance the experience of the visually impaired museum visitor.

The comparison between studies that have used different modalities of stimulus presentation suggests that haptic exploration (i.e. active touch) typically results in significantly better memory than passive tactile stimulation on the body surface (cf. Gallace et al., 2008). The results of the studies reviewed here clearly suggest that tactile information might be successfully used to provide different kinds of educational information to children and infants (see Minogue and Jones 2006, for a recent review of the important role that haptic stimulation might play in education). Interestingly, the twentieth-century Italian educational movement of Montessori schools already highlighted the importance of touch (and, more generally, of 'sensory' education) in enhancing the learning experiences of children in schools (e.g. Montessori 1943). In particular, young children have to follow the edges of three dimensional letters with the tip of their fingers to improve their reading and writing abilities (see Montessori 1991).[5]

Following on from these arguments it would seem plausible to suggest that the information that is gained through multisensory stimulation (stimulation that includes the sense of touch) may provide stronger and longer-lasting memories for that information than for information acquired solely by visual or auditory stimulation (cf. Dinh et al. 1999; Farkas 2003; Hoffman et al. 1998; see also Murray, Foxe and Wylie 2005, for neuroimaging evidence suggesting that in a memory task the brain implicitly stores multisensory experiences even when they are not necessarily of any particular behavioural relevance to the task at hand). Note that the possibility of providing educational information via the sense of touch assumes even greater importance in visually impaired individuals. With regard to this topic, the present review has shown that haptic memory in blind individuals compares favourably to the visual memory of visually sighted individuals and that it might require shorter presentation times (as compared to the time required by sighted individuals) in order for the material presented to be retained effectively (e.g. Arnold and Helron 2002).

In conclusion, this review has demonstrated that tactile memory represents an important, if currently under-investigated, topic. In the future, cognitive neuroscientists and cognitive psychologists will need to further investigate tactile memory in more ecologically valid contexts, involving the retention of tactile information over longer intervals (i.e. months and years). The results of this research will hopefully lead to a better and more extensive use of tactile information in educational contexts, such as in museums and galleries.

Suggestions for the Use of Touch in Museums and Heritage Contexts Emerging from the Cognitive Science Research on Tactile Memory Reviewed Here

- The active manipulation of objects is preferable to passive tactile stimulation (cf. Gallace et al., 2008)
- Whenever possible, three-dimensional objects should be used to present information, rather than two-dimensional raised line drawings (e.g. see Klatzky et al. 1985).
- The presentation of tactile distractors should be avoided when new tactile/haptic information is being presented, while other forms of distraction (e.g. visual and auditory) should be avoided at later stages as they may interfere with a person's ability to retain tactile information (e.g. see Manning 1980; Watkins and Watkins 1974).
- It takes significantly longer to explore tactile as compared to visual displays (this is especially true for sighted individuals, e.g. see Berla 1982).
- Tactile stimuli may provide a particularly effective aid to improving the learning of younger people (e.g. see Catherwood 1993; Minogue and Jones 2006; Montessori 1943).
- Adding tactile pictorial stimulation to acoustic and verbal information can facilitate the learning and memory of visually impaired individuals (e.g. see Pring and Rusted 1985).
- Consistent multisensory stimulation regarding a given object/experience should be provided whenever possible (e.g. see Dinh et al. 1999; Farkas 2003; Hoffman et al. 1998; Murray et al. 2005).
- Whenever people need to inspect tactile stimuli that cannot be moved, then a canonical haptic orientation should, wherever possible, be used (see, for example, Lacey et al. 2007; Newell et al. 2005).

Notes

1. Note also that very little research has been conducted on the topic of olfactory and/or gustatory memory either (though see Annett 1996; Engen 1987; Rabin and Cain 1984, for exceptions).
2. Note though that although significant, this represents a much more modest improvement in participants' performance relative to the improvement seen in visual studies, where up to between 12 and 14 more letters can be reported when a partial (rather than whole) report procedure is used.

3. In both conditions of stimulus presentation, the participants were explicitly instructed not to look at their arms in order to rule out the possibility that the position of the stimulus presented on the body was coded and stored simply by means of visual rather than tactile information.
4. Note, however, that this activation in the visual cortex has been shown not only in blind participants but also in sighted participants trained to perceive tactile stimuli. This would seem to suggest that the reorganization of the visual cortex that has been shown in blind individuals might be, at least in part, the results of intensive practice in tactile discrimination (e.g. Saito et al. 2007).
5. See also Lillard and Else-Quest (2006) for recent research, published in *Science*, showing that Montessori students proved to be significantly better in reading and mathematics, executive functions, social abilities and creativity as measured by a series of test than the non-Montessori children).

References

Amedi, A., Raz, N., Pianka, P., Malach, R. and Zohary, E. (2003), 'Early "visual" cortex activation correlates with superior verbal memory performance in the blind, *Nature Neuroscience* 6: 758–66.

Anderson, J.R. (1987), 'Skill acquisition: Compilation of weak-method problem solutions', *Psychological Review* 94: 192–210.

Annett, J.M. (1996), 'Olfactory memory: A case study in cognitive psychology', *Journal of Psychology* 130: 309–19.

Arnold, P. and Heiron, K. (2002), 'Tactile memory of deaf-blind adults on four tasks', *Scandinavian Journal of Psychology* 43: 73–9.

Atkinson, R.C. and Shiffrin, R.M. (1968), 'Human memory: A proposed system and its control processes', in K.W. Spence and J.T. Spence (eds), *The Psychology of Learning and Motivation*, Vol. 2, London: Academic Press, pp. 89–95.

Baddeley, A.D. and Hull, A. (1979), 'Prefix and suffix effects: Do they have a common basis?, *Journal of Verbal Learning and Verbal Behavior* 18: 129–40.

Berla, E.P. (1982), 'Haptic perception of tangible graphic displays', in W. Schiff and E. Foulke (ed.), *Tactual Perception: A Sourcebook*, Cambridge: Cambridge University Press.

Bliss, I. and Hämäläinen, H. (2005), 'Different working memory capacity in normal young adults for visual and tactile letter recognition task', *Scandinavian Journal of Psychology* 46: 247–51.

Bliss, J.C., Hewitt, D.V., Crane, P.K., Mansfield, P.K. and Townsend, J.T. (1966), 'Information available in brief tactile presentations', *Perception and Psychophysics* 1: 273–83.

Bonda, E., Petrides, M. and Evans, A. (1996), 'Neural systems for tactual memories', *Journal of Neurophysiology* 75: 1730–37.

Brang, D. and Ramachandran, V.S. (2007), 'Tactile textures evoke specific emotions: A new form of synesthesia', *Abstracts of the Psychonomic Society* 12: 101–2.

Brooks, D. N. and Baddeley, A. D. (1976), 'What can amnesic patients learn?', *Neuropsychologia*, 14: 111–29.

Buchel, C., Price, C.J., Frackowiak, R.S.J. and Friston, K.J. (1998), 'Different activation patterns in the visual cortex of late and congenitally blind subjects', *Brain* 121: 409–19.

Candlin, F. (2003), 'Blindness, art and exclusion in museums and galleries', *Journal of Art and Design Education* 22: 100–10.

Catherwood, D. (1993), 'The robustness of infant haptic memory: Testing its capacity to withstand delay and haptic interference', *Child Development* 64: 702–10.

Coltheart, M. (1980), 'Iconic memory and visible persistence', *Perception and Psychophysics* 27: 183–228.

Cowan, N. (1984), 'On short and long auditory stores', *Psychological Bulletin* 96: 341–70.

Darwin, C., Turvey, M.T. and Crowder, R.G. (1972) 'An analogue of the Sperling partial report procedure: Evidence for brief auditory storage', *Cognitive Psychology* 3: 255–67.

Dinh, H.Q., Walker, N., Hodges, L.F., Song, C. and Kobayashi, A. (1999), 'Evaluating the importance of multi-sensory input on memory and the sense of presence in virtual environments', *Proceedings of IEEE Virtual Reality Conference 1999*, Houston, TX, 13–17 March, pp. 222–8.

Easton, R.D., Srinivas, K. and Greene, A.J. (1997), 'Do vision and haptics share common representations? Implicit and explicit memory within and between modalities', *Journal of Experimental Psychology: Learning, Memory and Cognition* 23: 153–63.

Ebbinghaus, H. (1885–1913), *Memory: A Contribution to Experimental Psychology*, New York: Dover.

Engen, T. (1987), 'Remembering odors and their names', *American Scientist* 75: 497–503.

Esteves, J.E., Geake, J. and Spence, C. (submitted), 'Investigating multisensory integration in osteopathic clinical examination: An exploratory study' (manuscript submitted for publication).

Farkas, R. D. (2003), 'Effects of traditional versus learning-styles instructional methods in middle school students', *Journal of Educational Research* 97(1).

Field, B. (1975), *Hands on Museums: Partners in Learning*. New York: Educational Facilities Laboratories.

Field, T. (2001), *Touch*, Cambridge, MA: MIT Press.

Gallace, A. and Spence, C. (2008), 'The cognitive and neural correlates of "tactile consciousness": A multisensory perspective', *Consciousness and Cognition* 17: 370–407.

Gallace, A., Tan, H.Z. and Spence, C. (2006), 'Numerosity judgments in tactile perception', *Perception* 35: 247–66.

Gallace, A., Tan, H.Z. and Spence, C. (2007a), 'Do "mudsplashes" induce tactile change blindness?, *Perception and Psychophysics* 69 : 477–86.

Gallace, A., Tan, H.Z. and Spence, C. (2007b), 'The body surface as a communication system: The state of art after 50 years of research', *Presence: Teleoperators and Virtual Environments* 16: 655–76.

Gallace, A., Tan, H.Z., Haggard, P. and Spence, C. (2008), 'Short term memory for tactile stimuli', *Brain Research* 1190: 132–42.

Garvill, J. and Molander, B. (1977), 'A note on the relation between cross-modal transfer of learning and cross-modal matching', *Umea Psychological Reports, No 126*, University of Umea, Sweden.

Gilson, E.Q. and Baddeley, A.D. (1969), 'Tactile short-term memory', *Quarterly Journal of Experimental Psychology* 21: 180–4.

Gregory, R.L. (1967), 'Origin of eyes and brains', *Nature* 213: 369–72.

Harris, J.A., Harris, I.M. and Diamond, M.E. (2001), 'The topography of tactile working memory', *Journal of Neuroscience* 21: 8262–9.

Heller, M.A. (1989), 'Tactile memory in sighted and blind observers: The influence of orientation and rate of presentation', *Perception* 18: 121–33.

Heller, M.A., Calcaterra, J.A., Burson, L.L. and Tyler, L.A. (1996), 'Tactual picture identification by blind and sighted people: Effects of providing categorical information', *Perception and Psychophysics* 58: 310–23.

Hertenstein, M.J. and Campos, J.J. (2001), 'Emotion regulation via maternal touch', *Infancy* 2: 549–66.

Hertenstein, M.J., Keltner, D., App, B., Bulleit, B.A. and Jaskolka, A.R. (2006), 'Touch communicates distinct emotions', *Emotion* 6: 528–33.

Hoffman, H.G., Hollander, A., Schroder, K., Rousseau, S. and Furness, T.I. (1998), 'Physically touching and tasting virtual objects enhances the realism of virtual experiences', *Journal of Virtual Reality* 3: 226–34.

Hollins, M. (1989), *Understanding Blindness*, Hillsdale, NJ: Lawrence Erlbaum.

Johnson, C.L., Paivio, A.U. and Clark, J.M. (1989), 'Spatial and verbal abilities in children's cross-modal recognition: A dual-coding approach', *Canadian Journal of Psychology* 43: 397–412.

Kappers, A.M. (1999), 'Large systematic deviations in the haptic perception of parallelity', *Perception* 28: 1001–12.

Kappers, A.M. (2007), 'Haptic space processing. Allocentric and egocentric reference frames', *Canadian Journal of Experimental Psychology* 61: 208–18.

Kappers, A.M. and Koenderink, J.J. (1999), 'Haptic perception of spatial relations', *Perception* 28: 781–95.

Katz, D. (1989[1925]), '*The World of Touch*, in L.E. Krueger (trans.), Hillsdale, NJ: Lawrence Erlbaum.

Kerzel, D. (2001), 'Visual short-term memory is influenced by haptic perception', *Journal of Experimental Psychology: Learning, Memory and Cognition* 27: 1101–9.

Kilgour, A.R. and Lederman, S.J. (2002), 'Face recognition by hand', *Perception and Psychophysics* 64: 339–52.

Kiphart, M.J., Auday, B.C. and Cross, H. (1988), 'Short-term haptic memory for three-dimensional objects', *Perceptual and Motor Skills* 66: 79–91.

Klatzky, R.L., Lederman, S.J. and Metzger, V.A. (1985), 'Identifying objects by touch: An "expert system"', *Perception and Psychophysics* 37: 299–302.

Koran, J.J., Jr., Morrison, L., Lehman, J.R., Koran, M.L. and Gandara, L. (1984), 'Attention and curiosity in museums', *Journal of Research in Science Teaching* 21: 357–63.

Krantz, M. (1972), 'Haptic recognition of objects in children', *Journal of Genetic Psychology* 120: 1565–71.

Lacey, S. and Campbell, C. (2006), 'Mental representation in visual/haptic cross-modal memory: Evidence from interference effects', *Quarterly Journal of Experimental Psychology* 59: 361–76.

Lacey, S., Peters, A. and Sathian, K. (2007), 'Cross-modal object recognition is viewpoint-independent', *PLoS One* 2 (9): e890 17849019.

Lederman, S., Klatzky, R., Chataway, C. and Summers, C. (1990), 'Visual mediation and the haptic recognition of two-dimensional pictures of common objects', *Perception and Psychophysics* 47: 54–64.

Lehmann, S. and Murray, M.M. (2005), 'The role of multisensory memories in unisensory object discrimination', *Cognitive Brain Research* 24: 326–34.

Lillard, A. and Else-Quest, N. (2006), 'The early years: Evaluating Montessori education', *Science* 313: 1893–4.

Magee, L.E. and Kennedy, J.M. (1980), 'Exploring pictures tactually', *Nature* 283: 287–8.

Mahrer, P. and Miles, C. (1999), 'Memorial and strategic determinants of tactile recency', *Journal of Experimental Psychology: Learning, Memory and Cognition* 25: 630–43.

Mahrer, P. and Miles, C. (2002), 'Recognition memory for tactile sequences', *Memory* 10: 7–20.

Manning, S.K. (1978), 'The effects of interpolated interference on tactual and visual short-term memory', *American Journal of Psychology* 91: 445–59.

Manning, S.K. (1980), 'Tactual and visual alphanumeric suffix effects', *Quarterly Journal of Experimental Psychology* 32: 257–67.

McGlone, F., Vallbo, A.B., Loken, L. and Wessberg, J. (2007), 'Discriminative touch and emotional touch', *Canadian Journal of Experimental Psychology* 61: 173–83.

Melzack, R. and Eisenberg, H. (1968), 'Skin sensory afterglows', *Science* 159: 445–7.

Miles, C. and Borthwick, M. (1996), 'Tactile short-term memory revisited', *Memory* 4: 655–68.

Millar, S. (1972), 'Effects of interpolated tasks on latency and accuracy of intermodal and cross-modal shape recognition by children', *Journal of Experimental Psychology* 96: 170–5.

Millar, S. (1975), 'Effects of tactual and phonological similarity on the recall of Braille letters by blind children', *British Journal of Psychology* 66: 193–201.

Millar, S. (1999), 'Memory in touch', *Psicothema* 11: 747–67.

Minogue, J. and Jones, M.G. (2006), 'Haptics in education: Exploring an untapped sensory modality', *Review of Educational Research* 76: 317–48.

Montessori, M. (1943), *Come educare il potenziale umano* [How to educate the human potential], Milan: Garzanti.

Montessori, M. (1991), *Educazione per un mondo nuovo* [Education for a new world], Milan: Garzanti.

Morton, J. and Holloway, C.M. (1970), 'Absence of a cross-modal "suffix effect" in short-term memory', *Quarterly Journal of Experimental Psychology* 22: 167–76.

Murray, D.J., Ward, R. and Hockley, W.E. (1975), 'Tactile short-term memory in relation to the two point threshold', *Quarterly Journal of Experimental Psychology* 27: 303–12.

Murray, M.M., Foxe, J.J. and Wylie, G.R. (2005), 'The brain uses single-trial multisensory memories to discriminate without awareness', *Neuroimage* 27: 473–8.

Murray, M.M., Michel, C.M., Grave de Peralta, R., Ortigue, S., Brunet, D., Gonzalez Andino, S. and Schnider, A. (2004), 'Rapid discrimination of visual and multisensory memories revealed by electrical neuroimaging', *Neuroimage* 21: 125–35.

Neisser, U. (1967), *Cognitive Psychology*, New York: Meredith.

Newell, F.N., Woods, A.W., Mernagh, M. and Bulthoff, H.H. (2005), 'Visual, haptic and crossmodal recognition of scenes', *Experimental Brain Research* 161: 233–42.

Oppenheimer, F. (1972), 'The exploratorium: A playful museum combines perception and art and science education', *American Journal of Physics* 40: 978–84.

Overvliet, K.E., Smeets, J.B.J. and Brenner, E. (2007), 'Parallel and serial search in haptics', *Perception and Psychophysics* 69: 1059–69.

Pascual-Leone, A., Wassermann, E.M., Sadato, N. and Hallett, M. (1995), 'The role of reading activity on the modulation of motor cortical outputs to the reading hand in Braille readers', *Annals of Neurology* 38: 910–15.

Phillips, W.A. and Baddeley, A.D. (1971), 'Reaction time and short-term visual memory', *Psychonomic Science* 22: 73–4.

Pring, L. and Rusted, J. (1985), 'Pictures for the blind: An investigation of the influence of pictures on recall of text by blind children', *British Journal of Developmental Psychology* 3: 41–5.

Rabin, M.D. and Cain, W.S. (1984), 'Odor recognition: Familiarity, identifiability and encoding consistency', *Journal of Experimental Psychology: Learning Memory and Cognition*, 10: 316–25.

Ricciardi, E., Bonino, D., Gentili, C., Sani, L., Pietrini, P. and Vecchi, T. (2006), 'Neural correlates of spatial working memory in humans: A functional magnetic resonance imaging study comparing visual and tactile processes', *Neuroscience* 139: 339–49.

Rönnberg, J. (1995), 'Perceptual compensation in the deaf and blind: Myth or reality?, in R.A. Dixon and L. Bäckman (eds), *Compensating for Psychological Deficits and Declines: Managing Losses and Promoting Gains,* Mahwah, NJ: Lawrence Erlbaum, pp. 251–74.

Sadato, N., Pascual-Leone, A., Grafman, J., Deiber, M.P., Ibanez, V. and Hallett, M. (1998), 'Neural networks for Braille reading by the blind', *Brain* 121: 1213–29.

Saito, D.N., Okada, T., Honda, M., Yonekura, Y. and Sadato, N. (2007), 'Practice makes perfect: The neural substrates of tactile discrimination by Mah-Jong experts include the primary visual cortex', *BMC Neuroscience* 7: 79.

Schacter, D.L. (1987), 'Implicit memory: History and current status', *Journal of Experimental Psychology: Learning, Memory and Cognition* 13: 501–18.

Schurman, D.L., Bernstein, I.H. and Proctor, R.W. (1973), 'Modality-specific short-term storage for pressure', *Bulletin of the Psychonomic Society* 1: 71–4.

Spence, C. and Driver J. (eds), (2004), *Crossmodal Space and Crossmodal Attention*, Oxford: Oxford University Press.

Sperling, G. (1960), 'The information available in brief visual presentations', *Psychological Monographs* 74: 1–29.

Tassinari, G. and Campara, D. (1996), 'Consequences of covert orienting to non-informative stimuli of different modalities: A unitary mechanism?, *Neuropsychologia* 34: 235–45.

Taylor, M.M., Lederman, S. J. and Gibson, R.H. (1973), 'Tactual perception of texture', in E. Carterette and M. Friedman (eds), *Handbook of Perception*, Vol. 3, New York: Academic Press, pp. 251–72.

Walk, R.D. and Pick, H.L., Jr. (1981), *Intersensory Perception and Sensory Integration*, New York: Plenum.

Watkins, M.J. and Watkins, O.C. (1974), 'A tactile suffix effect', *Memory and Cognition* 5: 529–34.

Weinstein, S. (1968), 'Intensive and extensive aspects of tactile sensitivity as a function of body part, sex and laterality', in D.R. Kenshalo (ed.), *The Skin Senses*, Springfield, IL: C.C. Thomas, pp. 195–222.

–12–

Aesthetics of Touch among the Elderly

Michael Rowlands

I suggest in this chapter that we know our personal space principally through touch and, in particular, the *memory* of touch: of touching handles on doors, latches on windows, cutlery in drawers, plates in dark cupboards, and of dusting artefacts on mantelpieces and polishing furniture. If, through caring for objects, we care for ourselves and our intimate others, we do so, I argue, through the faculty of touch. To think of the experience of being at home is to realize how much depends on the familiarity that comes with knowing a space – by walking shoeless, by sleeping, by bathing, by preparing and serving food – and to grasp that domesticity is the sensuality of *being through touch* and *being in touch* with oneself and intimate others. Recently, for example, I counted the number of separate tactile actions needed to make a cup of tea – that is, from filling a kettle with water to disposing of a tea-bag and putting milk and tea in a cup; thirty five separate actions suggest an unconscious routine that turns a relatively small part of a space into a place of habit. One can replicate these movements into hundreds in an average day from getting dressed or undressed, brushing teeth and other ablutions, cooking, cleaning, and so forth. In other words, our daily experience of 'being at home' is largely a matter of following unconscious and repetitive routines of maintaining patterns of care for oneself and others, as well as the objects and materials that provision care. It is certainly understandable that whatever the need to 'move', involves adaptation to major changes in tactile habitual actions and an unconscious familiarity of a sense of place.

Introduction

The research project on which this chapter is based involved architects, town planners and anthropologists in understanding the material needs of the elderly who are 'in care'.[1] The ethnography involved a study of over 60 cases and a sub-set of 18 elderly people moving into cared environments in Harringay in North London. Each person was interviewed about the experience of moving from the autonomy of their own home into 'care'. The results consistently contradicted the assumption that the elderly would feel a loss of care for self and others as they moved to sheltered or cared environments. Assumptions about the moving of the elderly from domesticity to cared environments or sheltered accommodation – or, more crudely, about their being 'put in a home' – affect us all, directly and indirectly (and for most of us in a rather guilty fashion). The linearity of the phrase 'from domesticity to care' betrays the assumption that the move to cared environments involves a loss of the capacity to care for oneself and others, as in the frightening words 'taken into care'. Yet the institutionalized provision of care, whatever anxiety it invokes, is widely considered fundamental to a *humane* society.

A consequence of this tendency to associate care with institutionalization is to oppose domesticity to care in the various senses of the loss of the ability or the agency of caring, especially through the faculty of touch, both for oneself and others. The results of our research consistently contradicted the assumption that the elderly would feel a loss of care for self and others as they moved to sheltered or cared environments. If anything, the pattern was often quite the reverse. With increasing age, domesticity had become an unwelcome reminder of their growing incapacity to care, surrounded as they were by the material legacies of previous abilities to care: kitchens used now only in a minimal 'boiling-an-egg way', or spaces that could no longer be kept clean or functional. Unused spare bedrooms and difficult-to-maintain kitchens were an embarrassment, and caring for furniture and furnishings had become a burden. It was now beyond them to care for these reminders of former days of domestic utility and, perhaps, of grandeur. Among the informants there were, of course, nostalgic glances towards glory days of greater autonomy, and of being in control of their domestic circumstances, but consistently this was offset by statements of satisfaction with being cared for or not having to struggle with caring for oneself.

An example was Mrs S, who had lived in the same three-storey house in Stroud Green for fifty years; she had been married in 1942 and was widowed with three children and seven grandchildren. She was 80 years old and had developed Parkinson's disease, and her GP had arranged for her to move to sheltered accommodation two years earlier. How did Mrs S feel about the move?

> I was a bit upset at first but when I saw this flat it was the best thing that happened to me. Maureen [the warden] is a wonderful woman. But I was anxious. If the children hadn't been on at me, I don't think I would have moved.

Mrs S explained later that her anxiety related to her frailty as well as a concern that she would no longer remember where things were in her new environment. For instance, she often had to get up during the night to go to the bathroom and she needed to feel her way in what had to become an intimate space. When she was asked 'When you came here for the first time, how did you feel?' she replied: 'Well it was warm, the windows look out ... everything about the flat ... plenty of cupboard space ... I just felt free ... not being cold ... I felt free.' Which suggests that the old space had become a trial and the new warmer and lighter space allowed her to reinvent a sense of touch.

This discrepancy was not only a surprise to the investigators on the project; equally, it seems to have been a surprise for the elderly. Often, applying for sheltered accommodation had been seen as a humiliating matter, forced by circumstances and/or persuaded by friends and relatives, but, once done and dusted, it led to surprised experiences of renewed gaiety and life. To some extent, this was also a bit of a surprise for sons and daughters who may have been instrumental in the decision to sell the family home and to move mum or dad into a 'home'. They typically felt considerable guilt, and had difficulty accepting the surrender of their own responsibility for care of parents or grandparents because of personal circumstances. They had rationalized on the lines that having to be cared for is inevitable with age and it is the duty of the Council or Housing Association to provide that care.

The researchers had also anticipated a linear movement from a life of self-care and care for others in the domesticity of home (whatever that might mean) to one where the institution takes over and administers care. This may sometimes be the case, particularly with disability, and its implications for human dignity may be in fact, open to abuse. An extreme version of 'being taken care of' is being 'put into care'. Mrs T was vocal on this.

> An old people's home is where people go who've got nobody and they're not very agile ... like ... when they get older they get a little bit funny ... and they're not looking after [themselves]. They've got no relatives to go to, so they've got to have somewhere to live and someone to look after them and see that they wash themselves. They have a big communal lounge ... they all sit around in chairs and that ... they look at television and they look at each other. Then perhaps their warden or someone says 'Come on now girls, it's time for tea'. Then they all get out of their chairs and go and have tea. Then they all come back and sit in their chairs again. They all have their own little chairs – like children really, but they're old. I would never want to finish up my life like that. I would finish my life. I wouldn't ... couldn't live like that.

But the fact that the experience of the transition from one state of care to the other did not often match such assumptions made both by our informants and the researchers was a paradox to be explained. I argue that this is because we have not gone deeply enough into the question of care. We do not understand what it means to 'care' for someone or something. It is also important to recognize that, as with any good curator (in contrast to the *Oxford English Dictionary* definition of the term), this is done reflexively. Care is curation, and taking this out of the museum and into personal life involves understanding how decisions are made about what to keep and what to discard. Personal acts of curation, and the regimes of touch they require to be developed as habits, shape the process of producing a past and projecting a future for the self. Yet, the fact that the future of the elderly in Harringay was often seen by them in the very short term (five out of eighteen were over 90 years old), also meant explicit anxieties about getting that curatable past finally in order. New acquisitions for the future were low in priority but decisions on disposals were a real source of concern – though not in all cases. As we shall see, some cases (mostly male) fiercely resisted recognizing a stored past; instead, they anticipated events in their current life (visiting the pub, the betting shop, a friend) and focused on preserving continued personal access to public space. Either way, it is the metaphor of care as curation that requires further exploration.

Analogues can be drawn between personal acts of curation and the formal curatorial skills of documentation, classification and conservation of collections. Most of our informants, when asked, could remember the furniture and other objects that they had previously accumulated and in some cases given away or discarded. It's not that they kept a written record but, rather like the art of memory described by Yates (1966), they could remember a piece of furniture by its position in a room or by its use, and also by the person to whom they may have given it. A mental/visual classification of things once belonging to them was reinforced (if not sustained) by a reminiscence of touch – through the memory of having once used or cleaned or polished them. If one of the cardinal rules of the good curator is 'do not touch', the forms displayed by the elderly in Harringay are resolutely about who may touch. Preserving the right to say who can touch or use a thing is one of the more important priorities of disability; to say who can help move you, carry you, or bring things to you helps maintain a sense of self.

As researchers, we had taken the notion of care and caring environments rather for granted, which diverted our attention in the study to the objective circumstances of 'care'; how much space was provided, the functions of space, the use of objects, and the like. The literature on ageing is generally devoid of questions about the nature of care. In retrospect this is surprising, given the central importance of the ontology of care for the development of both phenomenological and existential philosophies in twentieth-century thought. In *Being and Time*, Heidegger recounts an ancient creation myth in which a deity, Cura, shapes human beings from clay (donated by Earth) and infuses them with spirit (donated by Jupiter), whereupon the beings created are named as human by Saturn (Mulhall 2005: 112–13). Heidegger uses this myth of the shaping of Being to argue that care is the basis of being human – that is, Cura is the creator of *Dasein* (being-in-the-world).

In evoking this tale of human creation through curation, Heidegger made a distinction between *caring for* as anticipatory and future-oriented, and the actions of *taking care of* something or somebody as reflexive and past – or care as intention contrasted to care as disposition. To *care for* something or somebody is anxiety ridden; however much you may care, your caring is contingent upon unpredictable future events. The human condition is fated to a self and to a world of other selves and objects about which one cannot but choose to be concerned. To *have taken care of* something or somebody is reflexive of past actions, characterized by acts of bringing together to ensure the conditions of collective well-being. Whilst both these aspects of care relate to creation (i.e. creation of self), their separation is unified by their temporality; or, rather, as modes of caring they represent experience of different temporalities in the present. As we shall see, the more nostalgic may contrast with the more forward-looking, although both temporal experiences can exist in a personal experience.

With age, the capacity to care for something or somebody in the future runs up against the limiting condition of awareness of one's own death. This suggests the elderly's experience of care will be a more reflexive, past-oriented concern. As persons capable of self-reflection, we are all curators in the general sense of curating a life – in the use of objects, photographs, clothes, souvenirs, and household furnishings as a means of narrating a life story. In this respect, the elderly in Harringay are exemplary, if diverse, curators who reflect on their earlier circumstances of having cared for persons and things and in the process make sense of a life well spent.

Four Cases, with Different Inflections, Illustrating this Tendency to Curate a Life Reflexively

Mrs C is a widow with three sons, who came to Britain from Jamaica in the 1950s to take up a nursing career. She still has furniture and crockery inherited from her parents, and what she describes as 'all the things that make a home', waiting in storage in Jamaica for her eventual (as she sees it) return. In fact, she is disabled with rheumatoid arthritis and moved two years ago to sheltered accommodation in Wood Green, where she is visited every week by one of her sons. She tells a sad story of being married to an alcoholic husband who sold all their belongings to feed his habit. She was left with nothing when he died and, now disabled, had relied on others to help furnish her new accommodation.

She described the furniture in her flat in terms of who had given it to her. The settee was the oldest. It was given to her by a nursing friend twenty years ago and, though it is now old, she keeps it for sentimental reasons. Other friends gave her a table and chairs, her nephew bought her a fridge, and her sister-in-law bought her a special recliner chair, around which her day revolves. She eats there off a tray, watches television, keeps the telephone there, and reads borrowed books and a magazine called *Friends* given her by a neighbour upstairs. She keeps a garden chair in the flat to remind her of the days when she had a garden, and likes the flat because it has a sunny sitting room. This gathering in of second-hand furniture and things, she associates with cosiness.

She has a large collection of soft toys about which she says, 'they remind me of gentleness, because they are so cuddly'. She looks after them, rearranging them on her bed every morning.

They also look after her: 'You see that little monkey on top of the TV, he sits there. I say to him if I am going out, you watch the place while I am away.' She keeps photos of her family, most of which she took herself: 'It just reminds me that I am not alone. I have got a family even though I live alone.' One of the photos is of herself as a nurse 'when I used to be a school nurse. I always keep it to remind me that I used to be helpful to others'. She also stores cushions in two of her cupboards. 'I don't need them myself,' she explained, 'but they might be useful to people who fall on hard times'.

It is not surprising to see the home of Mrs C as the summation of a life of a carer curated around the responsibilities of care. Her story, as one of caring for others and others for her, suggests a rather nostalgic present of having once cared for others that her disability now prevents her from doing. However, the cuddly toys are the more active part of her concern for care because they may be both cuddled and touched, and are responsible for keeping order in the flat while she is away. As far as she is concerned, the toys care for her as much as she cares for them. In other words, the soft toys have agency, even a therapeutic value, in providing company and allowing themselves to be touched and handled that is reminiscent of 'grooming behaviour'.

At the other extreme is Miss G, who lives in a block of sheltered accommodation in Muswell Hill and doesn't appear to care for her things at all. She is in her seventies, unmarried with no children, and with a brother who lives nearby in the same block. She took the accommodation as a furnished flat rented for her by the Council. Most of the furniture belongs to the landlord and, when asked if she saw it as her home, she replied: 'Well, no. It's a place where I live. You make what you can of it, don't you?' When asked about what she would see as improvements to the flat, however, Miss G provided one of the longest lists of detailed recommendations that we got for the architects. She had prepared a written list of recommendations for how to improve the ventilation, for sealed windows to keep the heat in, for better ways of cleaning surfaces, for the priority of a washing machine, and many other recommendations that would, she felt, improve the lot of those in rented sheltered accommodation in general. She was very concerned about the implications of all this for others, and that the research would lead to tangible results.

We then asked Miss G about the collection of books and music cassettes that she kept in the flat. They turned out to be the one personal collection that she was very committed to keeping. Amidst all of the landlord's furniture, her own music system and the bookcase were the items for which she indicated some sense of possession; a collection of books and music cassettes turned out to be of central importance. 'Basically, I like books and I've got rid of so many over the years that I don't think I'd like to get rid of my books – and music tapes. I think they are the two main things I'd want to keep … I've always been a reader and I've always liked music.' This is because she is involved with a local church that has a club meeting once a week and she provides the music for the occasion. She not only does the music programmes for the club members but she also uses her books to devise quizzes. Amidst all the things belonging to the landlord, the music system and the bookcase were the items for which she felt a strong sense of possession.

Although she is in her seventies, through these activities she denies she is old, speaking about the care she takes of the 'old folk' in the club and how she likes to see their old photographs and hear their old stories. This relates to the fact that she was the lastborn following two brothers and a sister, and that she had stayed at home looking after her parents until she was well into her forties. When her father died, Miss G moved with her mother to Blackpool (her mother's birthplace) and where she lived until her mother died. She subsequently moved back to London to be near her brother.

When asked what had happened to all the furnishings from her mother's house in Blackpool, she replied that she had just got rid of it all. This lack of concern for her own things seems to be an attitude shaped by a life committed to externalizing a sense of care to others and their requirements rather than care for herself. Unmarried and undomesticated, she maintains her freedom to care (or not) and to refuse the burden of possessions that require care. A history of not owning much was summed up as follows: 'I've never been in a situation where ... I lived at home until my late thirties. I mean, the family gradually got married and moved away and I was the only one left at home (with her mother). So basically, I've not had any need to have my own furniture. It's never been a big thing anyway, it doesn't bother me.'

The study so far suggests that among the elderly informants the institutionalization of their care has promoted a reflexive concern with the narration of personal histories. Mrs C and Miss G, for example, are both concerned with remembering that they have personal histories of taking care of others, and of this being reciprocated. Mrs C is more subjective in feeling that her own care of others in the past has been recognized and reciprocated in the help she has received more recently in her disabled state. Miss G is more objective, and externalizing, in continuing her personal trajectory of taking care of others whilst neither expecting, nor perhaps wanting, this to be reciprocated. This suggests that, by relieving them of the anxiety of care for themselves or others in the future, the institutionalization of care promotes this reflexive turn that makes the present a sort of summation of their personal pasts, experienced through either active reminiscence about the use and disposal of possessions or a kind of absolute forgetting of the relevance of such memories.

But other case studies suggest that it can be more complicated than personal accounts of well-curated lives. Take for example, Mrs T's concern with her present situation and relative lack of concern with her personal past:

> Well, I've had many lives. Different stages, you know. This is my last phase, this is. I've just got to accept it. It's a bit of a job to accept it, you know. I shall be 90 in May. I mean, that is quite a good age. But I don't like it, I don't like the idea at all. I know that every day could be my last – you know, you only live for one day. You know, like a butterfly.

Yet, for someone living her last day, Mrs T is resolutely 'caring for', and anxious about the future. Her days are spent 'caring for' objects, things, and attached persons, whose well-being directly impinges on her own: 'Every day I do bits and pieces here and there. I do bits of washing – odd bits of washing. I clean some cupboards – something like that – just to keep going, you know.'

The obstacles that prevent her being able to care in the future are resolutely confronted and solutions found, so that she can still act in the world. When we asked her to explain why she had complained about not being able to go out at weekends, we expected a depressing story of the frailty of age. Instead, it was how the ordeal for her to walk from her flat to the laundry had been overcome through the help of a carer, who takes her laundry down and gets it done for her. There is little sign in Mrs T, therefore, of fear of loss of the capacity to care, either with age or her sense of closeness to death. Instead, the daily round is a constant battle of getting things and people in order.

Mrs T has a new chair that cost £200. She has replaced the cover on it from her old chair that had developed a broken leg, so it was still familiar to her. But the acquisition of a new item of furniture has put the arrangement of her living room into question.

> *Mrs T*: I always sit in my chair, but I'm seriously thinking of changing the room round because I really need another chair here, a comfortable chair. The lady next door has two chairs here, armchairs, but she hasn't got a table. You'll see her furniture is entirely different from mine. It goes all around the room. But I think to myself – what if I put that table there?
>
> *Interviewer*: You mean near the window?
>
> *Mrs T*: Yes, I could sort of move the place round and move that chair over here where the table is. I could put the chair over there, and Chris could still watch his TV.

We hadn't heard about Chris up until then, and discover that he is a neighbour's son who comes to visit her and sits watching her TV. But this arrangement would create further problems and, as usual, when one item of furniture is changed it means the whole system is upset, which Mrs T finally realizes: 'I might put that chair over there. I could put it anywhere. Actually, I'd rather wait till Margaret comes this afternoon and I am going to ask her what she thinks.'

There is no reason to assume that the more anxious future sense of 'caring for' dissipates either with age or with the security created by living in the conditions of institutionalized care. The principle that you never surrender the wish to remain a carer for people and things in the future asserts a primacy of 'caring for' in the everyday life of Mrs T over the reflexive 'taking care of' of Mrs C and Miss G. From a personal perspective, while we might say in all these cases their 'being in their worlds' is unified by 'care', the differences are experienced as differences in temporality. The more nostalgic experience of Mrs C contrasts with the more resolute past present of Miss G, and both contrast with the present future sense of what is important for Mrs T.

The study could be criticized at this point for promoting a system of 'ideal types' as an analytic model to subsume the detailed particularity of these case studies. But then we come to Mr K. He is 75, unmarried, with no children. He has one sister living in Ireland and another who lives nearby. He was homeless before he came to the refuge, and it was his sister who brought him to where he is now. Yet his response to a question about his sister elicited a series of denials about his need for care:

> *Mr K*: Yes, she brought me here, but she hasn't got anything to do with me.
> *Interviewer*: You mean she is not your sister?
> *Mr K*: She is my sister but she is nothing to do with me.
> *Interviewer*: Do you have other relatives?
> *Mr K*: Yes, but I don't bother with them.
> *Interviewer*: Does your sister live near?
> *Mr K*: Oh, yes, at number 32. I see her every day. When I was ill, she cooked for me every day.
> *Interviewer*: Do you go out?
> *Mr K*: Well I go out to play music – Irish music.
> *Interviewer*: Do you play in a group?
> *Mr K*: I do not. I play on my own. It keeps me busy.
> *Interviewer*: Can I take some photographs of the flat?
> *Mr K*: You can, but I don't want to be too long now because I have to go out.

The interview was so short that it is probably inappropriate to say much, except that the image of Mr K's own personal 'lack of care' inside the flat connects well with his desire to be out of it – in a pub, playing music. What he was concerned with doing was outside, and his irritation with sources of constraint on his mobility was fairly palpable.

Shrines to Touch

The idea that old age is a time for downsizing, simplifying and streamlining holds impressive sway, and yet, there is evidence that post-retirement years are also characterized by an inclination to complicate everyday life, both spatially and temporally. Resistance to the idea that ageing corresponds with loss of social and emotional relations is found in the curating of displays of photographs, ornaments and pictures in special places for their display that we interpret as shrines. Mrs C, for example, has ranks of family photographs that appear to be the single most important category of things on display and she makes the observation: 'They just remind me that I am not alone – that I have got a family even though I live alone'. In a ritual sense, shrines are embodiments of the relations between devotees and deities or spirits. A shrine usually holds a relic or cult image which is the object of worship, offerings and/or treatment, so that certain desired effects may be achieved. Shrines must therefore be attended to and/or visited so that they will continue to be places in order that their images and reliquaries retain their power. Secular shrines have similar associations with dedication to persons or subjects whilst entailing ideas of remembrance or commemoration. The idea that the ritual power of shrines is additive, depending on the powerful accumulation of images and relics over time, also accounts for their complicated physical appearance. Shrines, like altars, may have complexes of images and objects but they usually form a pattern of association that relate the visual to the tactile in a prescribed manner to anticipate a beneficial response.

In the Harringay sample, the majority of women with families had arrangements of photographs, ornaments, candles and vases to hold artificial flowers that in some sense mirrored the idea of a shrine or altar as a sacred site. Mrs C, for example, arranged framed photos of herself as a married woman and those of her sons and grandchildren in an arc around a rather Christian idea of candle holders, vases for flowers and a central display of heart-shaped ornaments.

She pointed also to a picture of herself as a nurse which reminds her of days when she cared for others. Mrs T also had pictures of her daughter and two grandchildren arranged around two candles and a glass offering-type bowl where she said she would keep sweets or biscuits for when the children would come and visit. Another arrangement was to have porcelain figures and ornaments around a centrally placed photo of grandchildren.

As we might expect photos of the family are the central ‘reliquaries’ around which other ideas of value – souvenirs from holidays, a sense of expensive gifts and flowers are arranged. In a more minimalist sense, Mrs G has a single collage arrangement of photos on her wall with a picture of her mother in the centre. It hangs above the bookshelves where she also keeps her music system. Mr K, on the other hand, has no photos of family or friends, but on either side of a large clock on the wall of his sitting room are arranged pictures of his ‘heroes’ – Princess Diana, Jesus Christ, and two rock’n’roll stars.

When asked about the ‘shrine’ arrangement, invariably the response was to show concern that others who might visit should know they were family persons and they were not alone in the world. Even if, in some cases, they were rarely visited by their families, it was important that they were not denied their role as parents or grandparents. Tins of sweets or biscuits were kept ready for family visits, their photos were on the wall and presents were on display. It is the care of objects and images, and the arrangement of soft toys, flowers, photos and candles, that turns images and objects into shrines and, perhaps, will have the desired effect of bringing a visit by the devotees. Being cared for in this way may also make up for past lapses and gaps in one’s personal biography, considering the inevitable anxiety of care for children, family and friends in earlier life. Hence a certain amnesia is encouraged with the arrangement of objects and photos into a shrine-like arrangement that reconstitutes the self, with the potential to re-inscribe oneself in people’s memory as a more idealized subject than actually may be the case in everybody’s memory.

Shrines allow selective remembering around an idealization of kinship. Marcoux has described this as a process of auto-ancestralization; reiterating the anthropological point that death is a lengthy process, and later life becomes a process of making oneself a living ancestor (Marcoux 2001: 231). Using the 'journey' metaphor, he describes the whole process of moving (in this case to sheltered accommodation), as a means of reconstructing the self in old age and constructing one's own death. I would attribute the reflexive 'taking care of' in the case of Mrs C to such an ancestralizing process, as she is able to surround herself with a memoryscape of things that confirms her continued importance to others. The elaboration of shrines – invariably, the chair in which the incumbent sits much of the time takes a prominent position – has many of the hallmarks of a fetish-like urgency in which the totality of the image is capable of reaching out and motivating ancestral connections. In a sense, this care for objects is autonomous and self-constitutive, and the curative aspect lies precisely in the creation of a totality that is able to care both for itself and its ancestral being. Miss G, and her books and cassettes, is part of the process of doing this for others, the self-sacrifice demanded by ancestors. But others such as Mrs T or Mr K are more resolutely independent and refuse to construct themselves in death. Instead, care for them remains future-oriented. Their concern with objects and people is contingent and relative depending on convenience and use, which invariably is about being free to do something or go somewhere, no matter what that might be.

Conclusions

I started off with a deconstruction of the 'care' metaphor and its linear trajectory within the modern institutional care setting. The term 'care' is so identified with the institutionalization of care that the everyday concerns of people to care for, and to have had a past of having taken care of others, tends to be obscured. The idea of 'being taken care of' or being 'put into care', bequeaths images and metaphors of nineteenth-century death, poverty, the workhouse and urban isolation that, in fact, are not the experience of many in this sample – perhaps precisely because it is these images that have had to be worked against in institutional settings of care.

The examples discussed in this chapter suggest that care relates to different temporalities, often within the experience of the same person. Everyone, to some extent, has to 'care for' and to 'take care of', and expects this of others. Who or what might be the object of care, or who or what cares for you, need not matter. But hard lives can lead to reflections on the past and an increasing emphasis on summing up a past of care for families and others that, in many complicated ways, may help to justify a life and make it complete. The need to do this is no doubt a feature of age and anticipation of one's own death, so that one cannot experience or 'care for' or expect to be 'taken care of' except in a transcendental manner. This may or may not be according to personal religious disposition, but, in the broader sense of 'religion', literally meaning 'to bind', the ethical imperative that insists that it is human to 'care' is strongly metaphysical.

How may a curated life contribute to achieving this? We can recognize how the institutionalization of care shapes these personal dispositions. Curating a life as a kind of 'repair job', that serves to forget the failed bits and create a purer narrative of care for others, is facilitated by the cared environment that finally relieves one of the obligation to sacrifice. In the face of one's own death being not far away, the need to reflect on a life well led is a special kind of memory work that places particular value on personal ideas of a successful life. The transformation into a 'living ancestor', venerated as a holder of memory, creates shrines of material culture that family and friends may visit in a state of some reverence. Yet a similar facility of the provision of institutionalized care promotes the continued life of a renouncer, able to break with the past and take advantage of change. Here, the will to be free both rejects the past and also displays irritation with those who wish to ancestralize. One can equally imagine the irritation of children and others who find the elderly irascible and unwilling to play their role. The idea that 'things' do not belong to one, or can be changed or disposed of at will, has the added benefit that visitors may never know quite what to expect. This idea that there is nothing fixed about cared environments, whatever the assumptions behind their provision, aptly summarizes this need to care for objects, and the feeling that objects care for us in the curation of personal lives.

Note

1. The project was entitled 'From Domesticity to Care: A Study of the Mobility of the Elderly in UK' and was funded by EPSRC. It was carried out at University College London between 1999 and 2001, led by Juliet Hanson of the Bartlett School of Architecture and Town Planning. The study focused specifically on the material provisioning offered by local councils to the elderly in seven different parts of the UK. Whilst the major responsibility lay with the UCL Bartlett School, it is to their credit that they provided for an ethnographic component. I am grateful to my anthropological colleagues, Leone Kellaher and Jean Sebastian Marcoux, for their help and collaboration in writing up the project report on which this account is based.

References

Marcoux, J.S. (2001), 'The Casser Maison ritual', *Journal of Material Culture* 6(2): 213–36.

Mulhall, S. (2005), *Heidegger and Being and Tim*, London: Routledge.

Yates, F. (1966), *The Art of Memory*, London: Routledge and Kegan Paul.

–13–

Reminiscence: Recent Work at the British Museum

Laura Phillips

In order to build upon previous experience, reach out to local communities and generate knowledge for the development of future programming for older adults, a pilot project was recently completed by the Department of Coins and Medals within the British Museum. The project used the department's handling collection as a focus for discussion, learning, reminiscence and socializing. The project revealed how valuable objects were for stimulating participants, encouraging their input and enabling access to collections, both intellectually and physically. Both recognizable and unfamiliar objects were selected in order to explore the power of objects to evoke memories and compare this with the value of enabling older adults to engage with new subject matter removed from their own remembered pasts. One of the aims of the project was to investigate how object handling might meet some of the needs of older adults and improve their well-being. By increasing the confidence and skills of participants, the project is planned to continue through the loan of objects for use independently.

Introduction

The Department of Coins and Medals within the British Museum holds collections that span the last 3,000 years of human history and come from across the globe. The department's collections number around three-quarters of a million objects relating to the creation and use of coins and other forms of money and medals.

As a curatorial department, the Department of Coins and Medals is perhaps unusual, as it not only works closely with the museum's learning and interpretation specialists, but also, with support from HSBC, employs a member of staff to develop the interpretation of its internationally renowned collections. This includes the development and management of the HSBC Money Gallery and of public learning programmes both on and off the museum site. This programming centres on objects from the collections including a handling collection. It is through the use of a handling collection, both at the museum and during outreach, that visitors can get up close and personal with objects.

The British Museum has been exploring the value of touch for visitors for a number of years. Since 2001, visitors have been able to handle museum objects on desks in the museum's galleries under the supervision and expertise of trained volunteers. In this way, the museum takes a calculated risk with selected objects, balancing robust collections care with the value of greater access for visitors. The first objects made available were from the numismatic collections, perhaps because many of these objects had been handled profusely whilst in circulation. The patina built up partly by touch on the surfaces of these objects adds to their authenticity and is not something that the museum's conservators remove.

The motivation for both the museum and the Department of Coins and Medals to do this kind of work is to reach out to local audiences, encourage visits and engagement with collections and remove barriers to interaction with the collections. By touching objects that had been held in the palms of contemporaries, a physical connection is formed with the past, leading to a deeper understanding of the period, the object and its context. After all, how many of the museum's curators would examine numismatic material without touching it, particularly as weight is so important in the research of many of these objects?

Influenced by the successful 'Hands On' programme of object handling in the galleries, and informed by research into their needs (Stevens 2006), a pilot project was developed with the aim of informing cross-museum delivery for older adults. The project was devised to extend the reach of the museum to older adults who might face barriers when attempting a museum visit. It was intended to create a model for sustainable outreach utilizing the numismatic collections. Throughout the project, the participants were encouraged to lead the exploration of the collections in the hope that their confidence and skills in handling and researching objects would build to a point where they could continue the project independently with the museum providing loaned objects for discussion. With this aim in mind, the Older Peoples' Activity Worker from the local Community Association (which partnered the museum on the project) was present throughout.

The project was conducted at the Community Association's premises, where older adults meet for lunch and activities. An eight-week programme of reminiscence and object handling was developed with group members aged between 70 and 95 years old who had a variety of access needs, including requirements surrounding mobility. All participants had the opportunity to visit the museum and the project was allowed to evolve, with participants having input into the objects they would like to see, the topics they wanted to cover and the galleries in the museum they would like to visit.

It was hoped that the project would show that utilizing museum objects could be a valuable tool for engaging older adults, encouraging discussion and socializing. In order to allow the project to have a personal connection to the participants, familiar objects were used which could spark reminiscences and might enable the group to get to know one another and perhaps feel unity through common experiences. Through participants listening to each other's memories and opinions, it was hoped the project would encourage all the participants to feel valued. Reminiscence has the potential to encourage dialogue, both between participants and between the participants and the museum, allowing the latter to learn from the former through their interpretation and memories of objects from the museum's collections. The project aimed to build on previous experience of visitors providing valuable contextual information for collections from the remembered past, but with a strong desire that the benefits to the museum should be secondary to those for the participants. The project aimed to find out more about the kinds of provision older adults wanted and required and how the museum might meet their needs.

There were further questions and concerns that the project wished to explore. If object handling could evoke powerful memories, might they be difficult or negative memories? Were there safe topics of discussion or even safe objects which could be trusted not to cause harm? Might it be therapeutic for an individual to work through difficult memories in a safe environment and, if so, could a museum provide such an environment? Could a museum worker provide the necessary support? Would the voluntary nature of reminiscence protect participants? Would it be good for participants to have the opportunity for risk-taking by trying something different? The breadth of the collections enabled the project to extend further than the remembered past and beyond the boundaries of the United Kingdom. Objects that museum staff members expected the group to find unfamiliar were included with the intention of stimulating creative thought and, due to the concerns outlined above, ensuring that the project did not dwell only on the past.

The Power of Objects

During the project, the extraordinary power of objects could be observed through the participants' amazement at handling and discussing a 2,500-year-old Athenian tetradrachm, the recognition of a ten-shilling note and the surprise when participants realized something they had never known about the pound coin in their pocket. During evaluation, the objects were thought to be the 'most enjoyable' element of the project. This was consistent in all the evaluation interviews but for a variety of reasons: they evoked the past, they were interesting, they were important and significant and yet the participants were allowed to handle them, they encouraged comparison between now and then, they were a conversation and socializing tool, and the participants learnt new things through their inclusion in the project. Comments included 'I didn't realize there was so much about them', they promoted 'thinking about things you never thought of before', and were 'something to talk about'. Participants were often surprised how much they were enjoying a project based around numismatic material and all wanted the project to continue long-term. Observation indicated that the participants had bonded as a group, often happily chatting together before sessions started and over lunch once the sessions were over.

When recalling memories, object handling seemed to bring more vivid recollections in comparison to the use of images. For example, a detailed recollection of one of the participant's fathers bringing a large, crisp, white five-pound note home in his wages had been the cause of much excitement. The fiver was carefully folded in his top pocket and the children were made to wash their hands before they could touch it. Museum staff offered to source images for the participants to use in collages or in discussion, but it was the physical objects that could be brought to the group which were most likely to have their images included in the participant's artworks. This is interesting as, if participants had been asked prior to the objects being handled what kinds of images they thought they might put into a collage, it is unlikely that images of money would have been top of their lists.

Objects were used to spark reminiscence, but also diverse discussions such as the histories of the objects, their life in the museum, conservation issues, their relevance to the present, and so on. Objects could provoke very varied responses: cowrie shells provoked discussion of their use as money, the countries in which they were used, summer holidays in Britain, travelling abroad, jewellery and service overseas. Pre-decimal coinage provoked discussion on pocket money and what was bought with it, wages (the participants own and their parents), working life, the differences in the cost of living today with times gone by, leisure activities – to name but a few. The objects were almost always central to the discussion. With less recognizable objects, background information was sometimes given first, but often the discussion was simply steered and complemented with snippets of information that the member of museum staff felt might be interesting, such as Mary Gillick being 71 years of age when she designed the portrait for Elizabeth II's first set of coins. Objects allowed the role of the specialist to dissolve. In many cases, the participants led the inquiry and made the deductions, and the discussion was much enriched by their skills and experience.

As well as encouraging communication and engaging the participants, the objects enabled different levels of interaction so that individuals could get involved at the level they felt most comfortable with: they could just handle and think, they could talk to their neighbour or perhaps to a friend or family member after the sessions, they could say something factual about something they recognized or just state they remembered it, they could have a stab at guessing what an unfamiliar object was, what is was used for, where it had originated, and they could ask a question or allow the object to prompt the sharing of a memory.

This meant that reminiscence was just part of the project. The flexibility brought by the objects effectively allowed three members of the group to talk about their personal histories, whilst two others enjoyed discussing the objects, but less so their own lives. The objects, therefore, made the project more accessible for a variety of participants, enabling them to choose the level of involvement they wanted. The potential of objects to stir an unhappy memory could not be avoided, however, and no object was 'safe' due to the multiple associations provoked. In a reversal of this, the objects for one participant seemed to stimulate a high level of nostalgia for a past, happier time. It was difficult to engage this participant in positive discussion of the present and instead she was pleased to be reconnected to objects from her past. This participant's reactions revealed that she needed support in order to enjoy and value her present. This experience indicated that, although there is the potential for reminiscence to provoke undesirable reactions, perhaps these reactions could be read in order to reveal participants' care needs and so shape and target the response.

One area where the object handling had a very positive effect was physical access. The element of touch improved access for one participant who was visually impaired. For him, general museum visits were limited in enjoyment and interest as he couldn't read the interpretation and he felt there was often no one to ask about the displays. During the project, he could use a magnifying glass freely, feel the surface of objects and bring them as close as he needed. Object handling provided a way for him to access the collections successfully.

As the project progressed, participants' confidence with objects rose, so that by the final session, the group, including the Older People's Activity Worker, discussed a number of unfamiliar objects for an hour without the input of a member of the museum staff. There was a discernable knowledge and skills increase, the group's discussion and research skills and their confidence to guess what an object might be and explain their reasoning to the group was much developed. This confidence was also discernable during the museum visit, when the group were happy to explore both together and individually, discussing displays with little reliance on museum staff. Group members remembered objects they'd seen seven weeks previously, reinforcing the idea that active learning can improve a person's retention of information.

Conclusion

In conclusion, objects were a valuable focus for discussion and reminiscence as they provided a talking point, stimulation and increased enjoyment for participants. Difficult memories and nostalgia could be triggered through reminiscence, and a positive conclusion to this could not be guaranteed. Museum staff and other professionals leading sessions must be aware of the power of evoking memories, and recognize the importance of seeking the advice and physical presence of a member of staff with more appropriate training and who knows the participants. Not only is this invaluable as the project is happening, but the staff member can provide sustained support past the end of a project. Participants' reactions to objects animated them and revealed something of their personalities – and, in some cases, their needs. In addition to helping a group of peers to bond with each other, this also illuminated a possible way forward in developing meaningful programming for older adults and their carers, particularly those in residential care. Carers could inform their work through reminiscence and object handling, by finding out about the lives and opinions of the older adults they assist. This could improve care by stimulating it to be more individual, respectful and dignified. Objects made the project more accessible physically through providing for different senses; but they were also beneficial intellectually, as one object could provide a trigger to a broad range of topics, both personal or not, allowing greater flexibility towards participants' needs and learning styles. Objects enabled a wide variety of interaction, encouraging diverse engagement and allowing all participants to be involved and to provide valued input. This project proved that utilizing a combination of objects – both recognizable and unfamiliar – enabled the participants to talk about more than just their memories. The objects promoted learning, creative thought, skills development and greater confidence in the participants. As one participant stated, 'objects stick with you; you can't get rid of this past giving a real sense of the power of objects when used in a project such as this'.

Acknowledgements

This project would not have been possible without the support and valued input of Helen Causer, Older People's Activity Worker, and that of all the participants at Millman Street Resource Centre, part of Holborn Community Association.

References

Monti, F. (2004), *Evaluation of the British museum 'Hands On' Programme*, London: British Museum (upublished paper).

Orna-Ornstein, J. (2001), *Development and Evaluation of the HSBC Money Gallery at the British Museum,* British Museum Occasional Paper 140, London: British Museum Press.

Stevens, C. (2006), *The British Museum Research Project with Camden Older People*, London: British Museum (unpublished paper).

–14–

Getting a Handle on the Past: The Use of Objects in Reminiscence Work

Bernie Arigho

The opportunity to handle, explore and experience objects is part of good practice in reminiscence work. This chapter will clarify what reminiscence work is and illustrate how the use of objects can perform many functions in reminiscence projects: helping to start projects off well, helping participants to find memories, make contributions and achieve their objectives, assisting successful group reminiscence, helping participants and groups to identify themes of interest and helping participants to find activities to take part in. Reminiscence work is person-led and not object-led. It is part of person-centred care. The objects are only important in so far as they relate to some aspect of remembered life-experience and act as a trigger for some form of positive engagement. Reminiscence work advocates a multisensory and inclusive approach, defining objects in the broadest meaning of anything presented to the senses or the mind. Reminiscence work may also use the imagination and creativity to recreate objects and experiences from the past.

Introduction

What is Reminiscence Work? The Work of Age Exchange

To reminisce is more than the ability to remember facts and figures and long lists of things. To reminisce is to recall, retrieve and recollect remembered experiences from one's life: the stories of life that help us to learn how to live and be, how to relate to the world, how to conduct oneself in life, and how to feel about life. Having thought about it, one may then go on to shape these reminiscences into forms that can be shared with other people – for example, in a spoken story, a written poem, a painting or a quilt. The word 'recollection' is one I like to use to describe what we are doing when we reminisce. A reminiscence is a 'recollection' – a gathering together of one's own remembered experiences and actions into a shape and order that tells a story, or paints a picture, of a personal episode or time in the past. The experience may be recollected in a vivid, detailed, narrative and sequential way, or it may be done more impressionistically and generally. It depends on what has been remembered, how it affects the person and how the person chooses to present it. It involves a kind of searching in the memory for experiences that can be put to some kind of use in the present and future. What kind of experiences, and what uses? And what helps with this process?

The subject matter of reminiscence is some past event directly experienced and reacted to by the individual, and brought to mind in the present by some trigger event. Once the event is brought to mind, there are various choices that can be made. You may want to persist in recalling, thinking and having feelings about the event, or you may want to try to forget about it and do something else. You may want to share the reminiscence with someone, or you may wish to keep it private. You may want it to be made public, or to be kept confidential. You may wish to express it in any number of creative art forms: a poem, a song or a dance.

So each person is like a great museum with their own personal collections of life experiences, some kept in mind, some put on display and some put to one side in storage for possible future use.

Age Exchange's role is to recognize this and to give people opportunities to recollect, share and 'put on display' their chosen reminiscences. This is what we call reminiscence work – 'the stimulation of social, creative and educational activities that value people's reminiscences'. Reminiscence work aims to help people reminisce well and achieve their goals for taking part in reminiscence projects across a range of personal, social, educational and creative dimensions (Gibson 1998). We focus on working with the oldest age groups, especially in care settings or with those who are excluded in some way, but in our experience the method is effective with all age groups.

Principles of Good Practice in Reminiscence Work

Phillips (this volume) discusses the use of coins to stimulate reminiscence and discussion. On the edge of British pound coins is inscribed *Decus et Tutamen* meaning 'an ornament and a safeguard', and this is how I regard the principles of good practice in reminiscence work. They help people's reminiscences to be richly embellished ornaments to life, and they provide safeguards to ensure that participants who share memories are respected and supported throughout the process. These principles have been developed by Age Exchange over the last twenty years through delivering more than 300 reminiscence projects in a variety of settings (see Table 14.1). A large part of our work at Age Exchange is to promote good practice and high standards in reminiscence work through a national training programme. This chapter will focus on the principle of using memory triggers that stimulate the six senses.

Table 14.1 Principles of Good Practice in Reminiscence Work (Arigho 2006)

The person-centred approach
Good communication – active listening
Genuine interest
Respect for personal choice
Fidelity and confidentiality
Establishing trust and rapport
Support for painful emotions
Non-judgemental attitude
Warmth
Good facilitation skills – equal opportunities
Use of memory triggers stimulating the six senses
Use of inclusive and relevant themes
A range of imaginative and creative opportunities
Monitoring and evaluation at every stage
Support, advice and guidance for fellow workers

The Use of Objects as Memory Triggers

Memories are triggered through the six senses of vision, hearing, touch, taste, smell and movement. There are a wide variety of ways in which people can be helped to find and choose memories to share. A reminiscence worker's collection of objects should include a wide selection of different types of images, sounds, textures, flavours, aromas and actions to go with the objects. Movement is the added ingredient, the sixth sense that is too often left out of the equation. We are stimulated to remember more when we re-enact an action or behaviour from our memories, such as a work routine, a game or a ritual of some kind. The objects were all made to be moved and manipulated in particular ways that were learned, memorized and embedded in the muscles.

A primarily verbal approach may work in some cases, when stories are ready-made to be told. When group members are new to the experience and uncertain about which memories to share, the non-verbal supplements to verbal triggers are a big help. When there are disabilities and verbal communication difficulties, a much more non-verbal and wide-ranging approach is required. We all have different abilities, strengths, preferences and experiences with regard to which of these senses is most effective in triggering our memories. A combined approach utilizing the senses that are strongest for people will maximize their opportunities for locating memories that they might want to share with others. Table 14.2 provides a checklist of different kinds of memory triggers that could be included in a reminiscence handling collection. The collection should be general and wide-ranging across a range of different reminiscence themes. It is good practice to include objects that are recognizable from all stages of the target age-group's life, so that participants are not restricted as to what times of their lives they will be stimulated to recall. It is entirely their choice as to what is of interest to them and why. Choice is the foundation stone of all good practice in reminiscence work. This involves giving people choices and then respecting the choices they make.

Table 14.2 Examples of memory triggers

Verbal	Key words related to people's interests, names of special people and places, specific technical language related to work, proverbs, mottos, catch-phrases, 'old' words, song lyrics, brand names, advertising slogans, poems
Non-verbal	
Visual	Photographs, postcards, newspaper cuttings, personal collectibles, film, slides, colours, birthday cards, scraps, old films, dreams, fashion magazines
Hearing	Music, film scores, recordings of children playing, transport, different kinds of work, birdsong, whistles, bells, musical instruments, the weather, street calls
Tactile	Animals, coins, people, sand, water, fabrics, artefacts, natural objects such as seashells and cones
Smell	Food, drink, flowers, herbs, perfumes, cleaning and polishing materials
Taste	Food and drink, rationing recipes, old-fashioned sweets, remedies
Movement	Dance, games, crafts, greetings and gestures, skating, cycling, different kinds of work

The Use of Objects to Clarify the Aims and Methods of Reminiscence Work

One of the stumbling blocks in trying to get a reminiscence project off the ground is the need to ensure that all those concerned understand what they are signing up for. It can be quite a new unfamiliar concept for some people and there may be some uncertainty or misconceptions. Reminiscence objects can be used from the outset to make it clear exactly what a reminiscence project involves. Using words alone to try and explain the process is not as effective as giving some practical examples of typical reminiscence activities based on good practice. Words alone are full of confusion and multiple interpretation. Words with appropriate actions are clearer and more informative. Some people may react quite negatively to the word 'reminiscence' and say that it is not something they would wish to do. These same people may then have memories triggered by an object they recognize, and begin reminiscing enjoyably and earnestly. 'Reminiscing' thus comes to mean 'expressing and enjoying oneself in the present' rather than 'dwelling on the past'.

Using objects as starting points also helps immediately to demonstrate some of the different ways in which people can benefit from taking part. Self-esteem can be expressed, common links can be found and new knowledge can be gained in the opening exchanges of an object-based and person-centred activity.

By making it clear what the process involves, we can obtain well-informed consent for taking part. Participants can be reassured as to the completely voluntary nature of their participation and the sharing of memories, and can gain the confidence to know that when they do choose to share they will be listened to and valued.

For many people it is helpful to make the beginning of the process more focused on external talking-points and conversation-pieces, such as the reminiscence objects, rather than on their own physical presence or internal thought processes. It may be easier and more comforting to engage with a concrete external object than to immediately search inwardly for what may ultimately arise as especially cherished or defining life experiences. The beginnings of any new group activity can be a tense occasion for some, and the more relaxed and involved we can help people to feel at this stage, the more they will be enabled to achieve through taking part.

The Use of Objects to Promote a Person-centred Approach

Reminiscence work is part of person-centred care, valuing people as individuals with unique life histories and life experiences and trying to see the past and the present from their perspective. There can be no predictions about what kind of objects will be suitable for any kind of person and there can be no assumptions about how people are going to respond to the objects. People respond in their own individual, unpredictable and interesting ways and then the reminiscence worker begins to get a better appreciation of them as individuals. There are no 'wrong' answers or responses to the objects. The objects become something different for each individual, stimulating different associations and connections arising from different life experiences. For example, a simple hand-woven basket at the start of a reminiscence group can generate many different individual reminiscences on a vast range of different themes. By passing it around the group and giving everyone a chance to say whatever it makes them think of, you may hear stories about apple picking, tea plantations, haggling in markets, queuing for rations, knitting, going to the beach, going to school, displaying flowers, visiting grandparents and carrying babies. This list was taken from just one reminiscence group at one meeting. There is no pressure to reminisce or make any contribution, just an open invitation for everyone to explore this interesting object together. It becomes clear that this is a group that values individual life experiences and diversity. The hand-woven basket is helpful because it is open, multi-purpose and multicultural and so means many things to many people. It opens up a whole world of possible contributions.

Sometimes an object is chosen because it has been mistaken for something else, or because it looks like something else, and the person is stimulated to share a memory about the object that has been called to mind. It can be helpful to have the 'wrong' object, as in not being the same object as in the memory can stimulate the imagination to recreate the 'right' object in the memory. The objects need only to be reminiscent – that is similar and engaging enough to stimulate reminiscence – for them to serve their purpose.

From these general beginnings, group members may be later prompted to think of some of their own personal objects and possessions with very specific personal associations. They may then wish to bring these to future group meetings and use them to assist their personal storytelling.

The Use of Objects to Assist Group Dynamics

Beyond enabling shared reminiscences, objects can assist with some of the basic requirements of any kind of group meeting. They help to make it clear whose turn it is to speak by giving the speaker the focus of attention. This in turn encourages the rest of the group to listen and pay attention, knowing that it will come to their turn and that they will then be afforded the same respect.

Skilled use of the objects achieves the twin goals of helping to reduce tension in the group, especially at the start of the project, whilst at the same time stimulating thoughts, memories and actions within the group. Group members are enabled to make choices about their involvement based on the stimulating options being offered.

Using the objects in creative ways for group activities helps to develop group-building and encourage interaction within the group. Such shared activities and tasks brings the group members together for a common purpose. (For many practical examples of how reminiscence objects have been used successfully in such ways, see Osborn 1993.)

Some object-based activities may become familiar and enjoyable ways for reminiscence groups to structure their sessions. For example, exploring an object such as the hand-woven basket has proved a good opening group activity. Playing games with, or demonstrating the use of, a range of objects are good ways of developing sessions. Listening to and singing along to music can be a good closing activity. In such ways the skilled use of objects helps to structure group sessions and so assist all stages of group development.

The Use of Objects to Identify Reminiscence Themes and Reminiscence Activities

One of the major tasks of the reminiscence group is to find its special theme(s) of interest. Reminiscence themes are big overlapping areas of human activity related to people, places, interests and times in one's life. Unless the group has come together to work on a previously agreed special subject, then finding the theme(s) becomes one of the opening objectives. Good reminiscence themes represent broad areas of human interest at all stages of life, no matter where you come from or how old you are. There are a number of universal themes that are of general interest to people of all ages and cultures – for example: food and drink, house and home, work, play, entertainment, travel, fashion, health, education, special people, special occasions and special places. As the project develops, it may become apparent that a narrower theme may interest the whole group. Examples of narrower themes include: cooking, housework, first job, childhood games, dancing, schooldays, seaside holidays, dressmaking, bringing up baby, love and friendship, weddings and my home town. A range of multisensory objects related to many different themes will help the group to find its theme(s), make links within the group and at the same time represent and appreciate the diversity within the group.

Another task for the group is to find the right activities through which their memories can be triggered and expressed. There are many different ways to express a memory or tell any kind of story. People come to a reminiscence project with a repertoire of their own skills, talents and preferred ideas for activities. The reminiscence objects are helpful in throwing up many ideas for activities based on the group members' own interests, skills and experiences. For example, a cotton reel alone has led on to the following creative reminiscence activities for different groups: French knitting, making 'toy tanks' and making kites. The objects lend themselves to movement and activity based on experience and knowledge. They would be sadly neglected if they were merely reduced to being talked about.

The Power of Memory and Imagination to Recreate Objects

The reminiscence groups may evoke memories of cherished or well-remembered objects that were in general use but are not in the handling collection, or that have been lost and can never be replaced because they were unique. The handling collection may help to identify these and it may be possible to locate and include some of such objects. One of the creative possibilities offered by reminiscence projects is the retrieval and recreation by the memory and by artistry of objects from the past. Creative activities that have assisted this process include miming a treasured personal object from the past, making an object using a traditional craft skill and re-enacting a memorable event in one's life. Thus, new reminiscence objects can be created out of the reminiscence group, following the creative and skilled use of an original 'starter' collection.

Conclusion

What is true of a reminiscence group is also true of a workshop in reminiscence work skills – namely, it works better with concrete examples of how to use reminiscence objects to stimulate memories, structure group sessions, identify reminiscence themes and develop creative activities. Age Exchange's training programme (www.age-exchange.org.uk) offers a range of courses focusing on different aspects of the work, concerned with promoting best practice in this specialist field. It is important to remember that the most effective memory triggers are people themselves, interacting respectfully, skilfully and imaginatively.

References

Arigho, B. (2006), *How To Help Reminiscing Go Well*, London: Age Exchange.
Gibson, F. (1998), *Reminiscence and Recall*, London: Age Concern.
Osborn, C. (1993), *The Reminiscence Handbook*, London: Age Exchange.

Part IV
Therapeutic Approaches to Touch

–15–

Enrichment Programmes in Hospitals: Using Museum Loan Boxes in University College London Hospitals

Guy Noble and *Helen Chatterjee*

This chapter describes the workings of University College London Hospitals NHS Foundation Trust Arts project: UCLH ARTS. It reflects on arts interventions within the hospitals and gives a brief synopsis of Arts in Health's beginnings and provides an overview of the burgeoning research into the field of Arts in Health. The chapter also examines a collaborative project with UCL (University College London) Museums and Collections: 'Using Museum Loan Boxes as enrichment activities for patients in University College Hospital (UCH)'. This project involved research undertaken by medical students as part of their student-selected component (SSC) and investigated the therapeutic or enrichment potential of taking museum loan boxes to patients' bedsides. The chapter outlines the aims of the project, its outcome and future challenges with particular reference to the variables in the project that need further research.

Introduction

The working environment of a hospital is obviously very different to that of a museum and or gallery, but in order to understand a little more about the role of the Arts in Health it is useful to describe the kind of working environment a hospital Arts Curator is faced with. This chapter will provide a snapshot of the UCLH Arts project and its interventions. It will then go on to describe the beginnings of the Arts in Health field and give an overview of the past and current research in the field. The chapter will conclude by examining a collaborative project with UCL Museums and Collections: 'Using Museum Loan Boxes as enrichment activities for patients in University College Hospital'.

The Working Environment of the UCLH Arts Curator

University College London Hospitals NHS Foundation Trust provides some 500,000 outpatient appointments and 75,000 inpatient or day case treatments a year, and it receives the equivalent number of visitors. With 6,000 staff spread across six hospital sites, it is one of the largest hospital trusts in Britain. The Trust serves anyone and everyone from the local, national and international community. The reasons for the public visiting a hospital are, not surprisingly, very different to the reasons of those people visiting a museum or gallery. First and foremost they visit for the services offered by the hospital – for treatment, advice or to visit a relative. Patients and the public visiting the hospital may well be in pain, stressed, confused and/or disorientated. The arts then are secondary in the hospital visitor's mind, but this does not mean the arts have to be second class; they can and do play a part in the service offered by the hospital. It is therefore the role of the Arts Curator within any hospital to provide a distraction from these worries and in so doing create a welcoming, uplifting environment for all patients, visitors and staff in the hospital through the use of a varied and stimulating arts programme.

What UCLH Arts Does

Exhibitions

UCLH Arts host exhibitions in the Street Gallery, one of the main thoroughfares of UCH, so-called because of its proximity to the Euston Road. There are some eight selling exhibitions annually where local community groups and artists can exhibit their work. Members of the public and staff are able to buy the work and a percentage of the sales go towards the running of the arts project. This has proved to be an excellent initiative within the hospital, as it not only encourages the local community into the hospital for a positive experience – and in so doing breaks down the perceived barriers between institution and community – but also it allows staff a chance to meet at the private views in a more informal environment, giving them a chance to speak with colleagues they would not normally meet up with.

Artist Residencies

Artist residencies bring patients and staff into first-hand contact with artists. Past residencies have included artist Frances Newman, who recorded the working life of behind-the-scenes hospital staff such as porters, cleaners, pharmacists and scientists. This residency celebrated the essential role of the staff and demonstrated the role of arts in helping to boost staff morale. Another residency involved a collaboration with a local gallery to bring textiles artists to young people being treated in UCH for chronic fatigue and cancer. Led by a number of different textile artists, the project allowed young people to explore and participate in a wide range of both traditional and cutting-edge textile techniques, materials, processes and art forms informed by contemporary textile art practice. This culminated in an exhibition of the participants' work in the Street Gallery. The physical act of making and engaging with the art had a tremendous beneficial effect on the young people's self-esteem and motivation. One participant, who sold his work to a local gallery, said 'I'm blown away. This has been one of the best experiences of my life!'

Music

Regular music concerts within the hospital's public areas, and smaller performances by patients' bedsides, have been a core part of the UCLH Arts programme. Music can have a huge impact on patients' well-being, reducing stress, anxiety and perceptions to pain (Staricoff 2004), and research backs this up together with extensive anecdotal evidence. A recent concert on a ward for older people at UCH showed how music can impact on patients' lives. Before the concert, patients were withdrawn, quiet and had no interest in interacting with one another. Following the performance of a singer-guitarist, the patients were less withdrawn and some sang along. The staff nurse on the ward reported that the patients were more upbeat as a result of the sessions and felt that the music had improved their hospital experience.

Site-specific Art Work

Site-specific art work forms the core of UCH's art collection. Funded through charitable sources the art works are integrated into the architecture of the hospital with a view to making the hospital environment more welcoming and engaging. Collaborative art works are also part of the site-specific commission programme and UCLH Arts has an on-going project with 'Workshop and Company', a mental health service for adults living in Westminster, and the 'Mosaic Workshop', a professional mosaic workshop, to create two site-specific mosaic art works for the X-ray department in the new UCH for patients, staff and visitors to enjoy. This unique project brings together artists, mental health service users and hospital staff to work collaboratively on this project and provides mental health service users and staff of the hospital the opportunity to take part in a number of workshops to learn about, design and create mosaics. The project can be seen to:

- promote understanding of mental ill health and the importance of social inclusion for people who experience mental health difficulties;
- develop the skills and confidence of mental health service users who take part;
- support the creation of new work for this unique context;
- foster creativity in the workplace by giving staff the opportunity to contribute to the making of mosaics;

- offer a cultural and meaningful distraction to patients and visitors from the stresses of being in hospital;
- enhance the recovery process through the use of art;
- strengthen links between hospitals and the cultural sector.

Origins of Arts in Health

The arts have been intrinsic to health since Ancient Greece, which can be traced back to the town of Epidaurus, in the Peloponnese. People from all over Ancient Greece used to visit the Temple of Asklepios and pray to the God of Healing in Epidaurus in the hope of a cure. As part of the healing process they would take some exercise and walk across to the amphitheatre where they could attend performances of tragedies or comedies (Senior and Croall 1993).

In medieval times, and on into the Renaissance period and beyond, caring for the sick (and poor) was the domain of the church. Physical illness was often seen as a manifestation of spiritual weakness, and hardly surprisingly within these buildings there were very often spiritual or devotional images on the walls and ceilings in an attempt to offer solace to the ill and heal the soul. Vast sums of money were spent on these important civic buildings and rich patrons came forward to pay for buildings and commissioned the great artists of the day. Examples of this can be seen in one of Europe's first hospitals Santa Maria della Scala, in Sienna.

It was not until the nineteenth century that a record of the effect of environment on patients was recorded by Florence Nightingale. From the much-quoted *Notes in Nursing*, Nightingale recorded that: 'The effect in sickness of beautiful objects, and especially of brilliancy of colour is hardly at all appreciated [by the authorities] … [The] variety of form and brilliance of colour in the objects presented to patients are actual means of recovery' (Nightingale 1859).

The past 25–30 years have seen a substantial growth in the Arts for Health field, from the establishment of hospital arts projects all over the country, to the formation of local and regional Arts in Health networks, such as the London Arts for Health Forum. It took the Department of Health until 2007, however, to publish a report setting out findings and recommendations on Arts in Health. This formed a joint publication with the Arts Council and the Department of Health, *A Prospectus for Arts and Health* (April 2007), in which the benefits of arts on health are laid out.

What are the Benefits?

Alongside the growth of arts projects in health care environments there has been a growing evidence base of the effectiveness of arts interventions in health-care environments and the impact this has on patient's health and well-being. The Arts Council publication *Arts in Health: A Review of the Medical Literature* highlights the impact that the arts have on health. This is a review of the published medical literature between 1990 and 2004 and focuses on a number of positive therapeutic and medical outcomes for patients, including:

- reduction of stress;
- reduction of depression and anxiety;
- reduced blood pressure;
- reduced pain intensity;
- reduced need for medication;
- improved mental health.

Despite all this evidence there is scant research relating to the use of museum collections on patient well-being. There is, however, some evidence that museum collections can play a role in the professional development of medical professionals. Research shows that first-year medical students taking part in art appreciation classes, which involved describing dermatological lesions, significantly improved their observational skills (Dolev et al. 2001).

Belkin and Neelon (1992) and Bardes et al. (2001) have found that the development of this crucial skill in medical practice has also been achieved by using different techniques, including learning to look at artefacts in an art museum. This lack of research into the use of museum collections in health-care environments led to Noble and Chatterjee's collaboration.

Using Museum Loan Boxes as an Enrichment Activity in University College Hospital

In 2006 Noble and Chatterjee set up a project to investigate the therapeutic and enrichment potential of taking museum loan boxes to patients' bedsides. They recruited medical students from University College Hospital to undertake this research as part of their student-selected component (SSC).

Introduced in 2003, SSCs are defined by the UK's General Medical Council as:

> that part of the course which goes beyond the limits of the core, that allows students to study in depth in areas of particular interest to them, that provides them with insights into scientific method and the discipline of research and that engenders an approach to medicine that is constantly questioning and self-critical. (General Medical Council 2003)

SSCs are a compulsory component in the first and second year of a medical degree and students are free to select a research project of their choosing from across the biomedical sciences.

One of the key reasons UCL Museums and Collections developed an SSC research project and embarked on this initiative was to involve museum objects in teaching and research in an innovative way in order to demonstrate the unique interdisciplinary role collections can play in university life. Importantly, the project sought to provide medical students with patient communication skills, methods of assessing well-being and research techniques. The core research aim to be investigated by the medical students was to evaluate the potential of object handling as a novel enrichment activity in hospitals. In doing this research the initiative also provided patients at UCH with a novel engagement activity and widened access to UCL Museums and Collections to a previously excluded audience.

The clinical basis for this project lies in the fundamental idea that improving the quality of the hospital environment and enriching the lives of patients while in hospital can have a positive effect for both patients and staff.

The aim of this research was to establish if museum collections and object handling could play a beneficial role for hospital patients, staff and students. We recruited five second-year SSC medical students who took loan boxes comprising objects from across UCL Museums and Collections (including art works, archaeological artefacts and natural history specimens) to patients' bedsides during the winter and spring academic terms of 2006/2007. Prior to data collection, students were given training in object handling, infection control, patient communication skills and data handling and analysis.

With training completed, the students began a series of visits to patients' bedsides at UCH in November 2006. The students began their sessions by asking several quantitative questions whereby patients completed a mood adjective checklist and two visual analogue or well-being scales. The former presented a series of moods (e.g. sluggish, nervous, vigorous, lively) with options to grade these emotions, from 'not at all' to 'extremely'. The well-being scales comprised linear metric scales with grades from 0 (e.g. worst imaginable health state and extremely dissatisfied with life) to 100 (e.g. best imaginable health state and extremely satisfied with life). Patients were required to grade their mood, their perception of their overall health status and provide an overall life satisfaction score at the start and end of each handling session (see Selai this volume, for a broader discussion of visual analogue scales).

The handling session involved the student first asking the patient to choose an object, and then a series of questions were posed as follows:

- Why have you chosen this object?
- What do you think this object is?
- What does the object feel like?
- Can you think of any experience that might relate to this object?
- Where do you think it comes from?
- What material do you think it is made out of?

Sessions were recorded using digital audio recorders, and at the end of the data collection period (March 2007) the students analysed the data using quantitative statistics and qualitative mind-mapping techniques to interpret the data and establish trends.

Outcomes

Quantitative analyses indicated an increase in overall well-being and patients' perceptions of their health status at the end of the session. Qualitative analyses revealed patients felt positive about the role of object-handling sessions as a distraction from everyday ward life and this had a positive impact on relationships amongst staff and patients. As a pilot research project, these outcomes are very informative and we are using these methods in parallel projects across the hospital to extend and continue this research.

A critical factor in the success of this project was engaging key people from the hospital, including clinical staff. The contribution of the Arts Curator (UCH) was crucial in engaging doctors, nurses and support staff to help with the training and the logistics of gaining access to patients. We also had to obtain the approval of the Research Ethics Committee prior to the start of the project.

The initiative demonstrates the value for medical student training in providing first-hand experience of patient contact in a context outside their clinical care. This gave the students a different perspective of patient care and an insight into patients' emotions while in hospital.

Importantly, the project has significantly raised the profile of UCL Museums and Collections within UCL by demonstrating its value as a core resource in teaching and research. Many university museums have faced challenges over the past fifty years or so regarding their role and value within higher education. With the many demands on student curriculum and resultant timetabling problems, the use of objects in teaching has been regarded by some, as having less core value. Many university museums and collections have experienced a downturn in their use at tertiary level. The initiative described above demonstrates that university museums must align themselves to the strategic priorities of their host institution if they are to become indispensable, and this involves developing teaching and research programmes that respond to the needs of the university audience. The teaching and research elements of this project illustrate that these collections can be useful tools for inspiring cross-disciplinary study and, at the same time, provide a novel research area for a new body of students.

The Potential of Object Handling for Women Facing a Gynaecological Cancer Diagnosis

Parallel research undertaken by research nurses at UCL Institute for Women's Health used a similar protocol to that described above in assessing the potential benefit of touch and object handling for women facing a gynaecological cancer diagnosis. The study involved women in hospital following surgery for gynaecological cancer and healthy women attending an outpatient gynaecological cancer-screening appointment. This preliminary research found that women in object handling sessions revealed significantly more information about their psychological state compared to standard counselling sessions. Further, that specific objects evoked deep emotional responses suggesting that some patients engaged with the objects on a symbolic level, perhaps revealing more than they would under normal circumstances. In such instances it was apparent that the object became the focus for uncovering emotions, acting as vehicle via which the patient felt more able to communicate intimate thoughts. Crucially, more research is warranted to evaluate and measure further the impact of object-focused interventions for people with cancer; nevertheless, there is enormous potential here for museum collections to play a significant role in therapy.

Conclusions

Critically, the results of this project demonstrate there is significant potential for museums and their handling collections to play a vital role in arts programmes addressing the needs of patients and staff. Further, that using a multimodal model involving patients, staff and students has exciting implications for the role of handling collections in student training and staff development.

Research into the therapeutic potential of object handling with hospital patients is ongoing, and future developments include a larger-scale clinical trial. This pilot research has demonstrated the enormous value of touch and object handling in a care setting, and we now need to characterize and understand fully the role and value of object handling for knowledge transfer and learning, skills acquisition, well-being, emotions and enjoyment.

Acknowledgements

We are extremely grateful to all of the patients at UCH and UCL Institute for Women's Health for their participation in this research. We are also grateful to our collaborators, Dr Usha Menon and Dr Anne Lanceley, who led the research undertaken at UCL Institute for Women's Health.

References

Bardes, C.L., Gillers, D. and Herman, A.E. (2001), 'Learning to look: Developing observational skills in an art museum', *Medical Education* 35(12): 1157–61.

Belkin, B.M. and Neelon, F.A. (1992), 'The art of observation: William Osler and the method of Zadig', *Annals of Internal Medicine* 116(10): 863–6.

Department of Health and Arts Council England (2007), *A Prospectus for Arts and Health.* Accessed online http://www.artscouncil.org.uk/publications/publication_detail.php?browse=recent&id=581.

Dolev, J.C., Friedlander, L.K. and Braverman, I. (2001), 'Use of fine art to enhance visual diagnostic skills', *Journal of the American Medical Association* 286(9): 1020–1.

General Medical Council (2003), *Tomorrow's Doctors*, London: GMC.

Nightingale, F. (1970[1859]) *Notes on Nursing: What It Is, and What It Is Not*, New Impression edn, New York: Dover Publications.

Senior, P. and Croall, J. (1993), *Helping to Heal: The Arts in Health Care*, London: Calouste Gulbenkian Foundation.

Staricoff, R.L. (2004), *Arts in Health: A Review of the Medical Literature*, Arts Council Research Report 36, London: Arts Council.

UCL Museums and Collections (2006), *Inspire: Highlights of UCL's Collections*, London: Grant Museum of Zoology, UCL.

–16–

See, Touch and Enjoy: Newham University Hospital's Nostalgia Room

Jackie O'Sullivan

Some three years ago following another musical afternoon on a care-of-the-elderly unit, it was realized how much patients wanted to reminisce and talk about their early years around the 1940s and 1950s, when most of them were still young and generally enjoying life. This realization led to the development of a unique and dedicated facility for patients, relatives, carers and clinical staff. This chapter discusses the development and success of Newham University Hospital's Nostalgia Room – a special facility created to encourage social interaction, distraction and relaxation away from the pressures of hospital life.

Introduction

In 2002–3 a small room in Newham University Hospital was transformed into a Nostalgia Room with a 1940/1950s theme. As a public space developed specifically for use by elderly inpatients at the hospital, the room was a response to concerns expressed by care-of-the-elderly staff at Newham Hospital about the isolation and despondency felt by many of their patients. The concept for the Nostalgia Room came initially from activities I organized in my role as Voluntary Services Manager when we engaged the services of musicians, arts and crafts professionals and dancers. These activities took the form of tea parties involving volunteers and nursing staff. Significant effort was put into these gatherings to make them as enjoyable as possible: table linen, bone china crockery and home-made cakes were the order of the day. In addition, old photographs of the local area were scattered across the tables to encourage interaction, reminiscence and discussion.

At first, the elderly patients seemed disinterested and failed to engage with the 'tea party' despite the volunteers' best efforts to encourage the patients to talk about their stories. One afternoon a breakthrough occurred thanks to the music of Ann Shelton (a 1950s sweetheart); a shy and introverted patient whispered to a volunteer 'that's Ann Shelton singing'. The volunteer suggested they sang together and so they both sang very softly 'A Nightingale Sang in Berkley Square'. The same afternoon another patient, who was admitted with malnourishment, ate two huge pieces of chocolate cake followed by two cups of tea, all rounded off with a rendition of the song 'Sweet Molly Malone'. These two patients improved very quickly and were both discharged home.

The hospital went on to organize more tea parties which always ended with singing or reminiscing about the 1940s and 1950s. It was due to the success of the musical afternoons that the idea was realized of creating a space capturing patients' stories and their past, and representing a time when they were still young and generally enjoying life. Following discussions with patients, relatives and carers, it was agreed that a space away from the wards, which was user-friendly, made patients feel relaxed, could be accessed easily and without formality, would potentially be of great benefit. With the success of the musical afternoons it soon became apparent that a Nostalgia Room could be the perfect solution to addressing the motivational and recreational needs of the elderly patients at Newham Hospital.

The Creation of the Nostalgia Room

The first step was to get the clinicians on side, as support was needed to obtain the room. Fortunately, the Director of Environment at the Hospital, who was also responsible for Arts within the Trust, gave his support by identifying an empty room and giving permission for a painter and carpenter to be available as and when required. Next a firm action plan was required. What would the room be used for and what would go in it?

The proceeding months became ones of seeking, hoarding, collecting, investigating, researching and acquiring. Materials and contents for the room were purchased, donated and loaned from a whole range of sources. Items were bought from boot sales, second-hand shops, auctions, and advertisements were placed for some items. The local Geffrye Museum was keen to become involved, to give talks and advice to volunteers; indeed some of the staff there donated their own personal possessions to the room. This was becoming a community project. The budget was tight, but we received a donation from The King's Fund, monies from the volunteers' fund-raising efforts and support from our own Hospital Trust.

The room was based on my memories as a child of the 1940s and 1950s. It had a piano, which was enjoyed most weekends when the mat was rolled back and 'knees-ups' were in full swing; the dining suite was the standard utility furniture and we had the exact match, except ours had had an iron scorch on the table where I had forgotten to pull the plug from the light socket! The television was sourced from the Internet and by chance was purchased from a retired doctor who had been a medical student with the hospital many years before; he insisted on delivering the set and, of course, we reminisced about how life was all those years ago for young doctors. There are many books, newspapers, comics, ration books, and even a book for the new bride giving advice on having her new husband's meal ready as he returns from work, with the new wife being freshly groomed, a clean pinafore, newly applied lipstick and a ready smile on her face to welcome the new head of the household!

The Aim of the Nostalgia Room

The ultimate aim of this initiative was to establish a room created in the period of the late 1940s, early 1950s, where inpatients from the care-of-the-elderly and rehabilitations wards could go for reminiscences, activities, social interaction, or just to relax away from the ward and clinical surroundings (Figure 16.1). The intention was to relieve boredom, motivate patients to socialize, let them relax, gain confidence, have self-worth, pride, to reduce the feelings of isolation, enjoy and share their stories and forget they are in hospital for a short while.

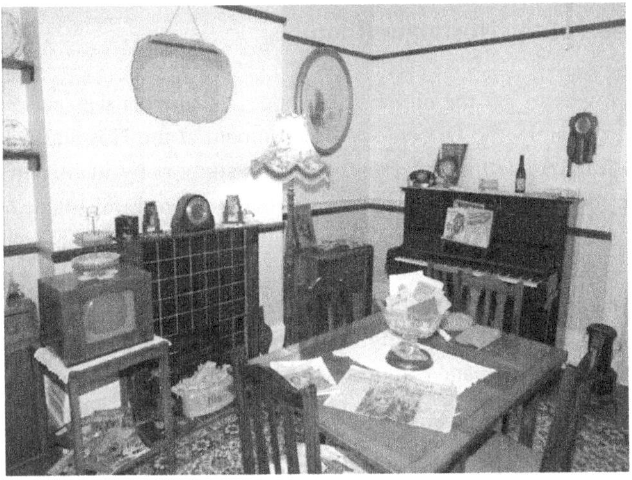

Figure 16.1 Newham University Hospital's Nostalgia Room.

The room has become a very hands-on place where patients pick up objects, reminisce and talk about them over a cup of tea. Piano playing with 'sing-alongs' are popular, as are other activities such as storytelling, painting and sketching sessions, sewing, Easter bonnet making, creating ceramic murals, playing draughts and cards. Artists, craft workers and others are often invited to run activities aimed at engaging patients, encouraging them to get involved and to provide a distraction from the stresses and worries of hospital life. The space creates a unique environment where patients, volunteers, relatives, carers and staff can realize the benefits of the role of arts in health. I believe this to be the first room of its kind with such authenticity and activities for patients and families to enjoy – and all the outcome of afternoon tea parties.

The following are just some of the many inspiring and thought-provoking stories demonstrating patient's responses to the room. The stories exemplify some of the important roles the room and its contents, the activities, and the staff and volunteers have played in the rehabilitation of the elderly patients at Newham Hospital. Furthermore, the stories highlight the valuable remedial implications of providing environmental enrichment for patients facing both physical and psychological challenges.

Case Study 1: Vi and her Companion

This story is about a patient who was engaged in a quilt-making session, an activity led by a textile artist who came twice a week and worked with patients to encourage them to tell their stories through the quilt by various textile means. Vi, a patient in this group, was crocheting and singing Vera Lynn's 'We'll Meet Again'. Sitting quietly was a patient with quite advanced dementia, her grandson and daughter by her side. This lady had no coherent conversation but she sang with Vi word for word. At the end of the song her grandson asked if we could do it again as he wanted his Granddad to hear his Nan singing, and so the song was repeated. This time the lady sang to her husband of many years – there wasn't a dry eye in the room.

Case Study 2: Ivy

Ivy, who had suffered a stroke, came to the Nostalgia Room on a regular basis and one day she was telling the volunteers that she had once owned a sewing machine, the same as the one in the room. Freda, a volunteer, asked Ivy to show her how to thread the needle, and under Ivy's instruction the deed was done. Next, it was suggested that Ivy have a go at sewing, but with the use of only one arm it was evident this would be challenging. With determination and help from Freda, who guided the material, Ivy did sew, turning the handle of the machine herself. By chance Ivy's daughter was at the door watching her Mum, and with tears streaming down her face, she said 'I am so happy to see my Mum doing normal everyday things again.'

Case Study 3: A Game of Draughts

The third story is of two men, one Asian and one a Londoner, both had suffered strokes and they were in the next beds to each other, but sadly had never spoken. Coincidentally both were brought along to the Nostalgia Room the same afternoon. After the preliminaries of looking around the room, one of the men picked up a board game and was asked if either had ever played draughts, both said they had and so – game on. The next day the nephew of the Asian man reported that his uncle had not stopped talking about the fun they had had and could he find out what the game was so he could buy it for his uncle. The upshot was the game was given to the two men who played continuously and became good buddies while in hospital, also encouraging other patients to join their games.

Case Study 4: A Vase of Flowers

Another occasion in the Nostalgia Room was a painting session when patients were asked if they would like to join in. One lady said 'I would like to paint but I don't think I could hold a brush now', but she came along anyway. This patient could no longer walk and her hands were very gnarled with arthritis, but nevertheless she decided to try and hold the brush to paint. The result was a beautiful painting of a vase of flowers. Everyone was extremely impressed and it transpired the patient had been a commercial artist during most of her working life. This delightful lady went on to paint all day most days and eventually some of her work was made into Christmas cards which the volunteers sold on their table sale during the festive season. On discharge this lady was given a set of watercolours and all the paraphernalia for her to continue painting in her new home.

Case Study 5: The Piano Player

A final brief story is about another lady who had suffered a stroke. This patient was starting to get movement back but had refused physiotherapy. She did, however, enjoy visiting the Nostalgia Room and playing the piano each day. This daily piano playing was all the physiotherapy the lady would agree to and eventually she regained all movement.

The Impact and Benefits of the Nostalgia Room

These brief stories demonstrate the impact a multisensory environmental enrichment initiative can have on patients experiencing a range of health concerns. In addition to real and measurable physical improvements brought about by interacting in the space, the room and the activities going on within it have demonstrable psychological implications for those involved. For both patients and staff this includes an opportunity to relax and enjoy the space and, through distraction, provides improvements in well-being and stress.

The Nostalgia Room has generated a huge amount of positive feedback. One consultant at Newham Hospital made this comment about elderly people who have a prolonged stay in hospital:

Occasionally there is the risk of becoming institutionalized and of developing feelings of depersonalization. Such factors can have a profound negative impact on physical, psychological and social well-being and as a result may hinder the rehabilitative process. It is paramount therefore that ways are found to maintain the dignity and self-worth of frail elderly people and this ground-breaking strategy helps to achieve this through the creation of the Nostalgia Room.

Some of the following anecdotes are from patients, carers and family members, inspired by the Nostalgia Room:

'Thank you for helping my mum know she can be happy again'
'This room makes me feel ten years old again'
'It's wonderful to hear my mum singing again'
'I could be in my granny's sitting room'
'May we come back after discharge?'
'When I am well enough, may I come back and volunteer?'
'Mum sewing, it's a miracle!'

The success of the room as a space which has significant remedial benefits has led to its use by a whole range of clinicians and other staff. Speech therapists take patients in the room when they want to encourage them to talk and receive therapy. Physiotherapists get patients to use the room to handle objects and so encourage dexterity. The Nostalgia Room forms a valuable part of the medical students' curriculum as a place where they gain an understanding of the importance of considering the whole person when treating a patient. In the wider context, local schoolchildren use the room to listen to the elderly people's stories, thus playing an important role in enhancing the pupils' knowledge of history.

Conclusion

The many positive outcomes of the Nostalgia Room can be seen on a daily basis in the faces of those who use and visit the room – the patients, their relatives, staff, volunteers and carers. The success of Newham University Hospital's Nostalgia Room is the 'see, touch and enjoy' philosophy which is shared by the volunteers, clinicians and all staff who use the room with patients, so that quality time is always available to the patients. This is an evolving room, and as the generations move on so will the room. Very soon we are hoping to engage a storywriter to capture the history and memories of our patients for future generations and to give schoolchildren an awareness of their heritage.

Critical success factors in this example of best practice include: the interactive nature of the resource, the involvement of specialist staff, clinicians, volunteers, relatives and patients in the construction of the space, community involvement and, finally, the support of senior members of hospital management. The Nostalgia Room experiences provide clear evidence of the powerful therapeutic potential of introducing heritage into health-care. The lessons learned here and at other organizations, including University College Hospital (see Noble and Chatterjee, this volume), demonstrate that there is considerable potential for closer links with heritage (including museum, libraries and archives), affording such institutions an opportunity to provide an important social service. These examples have but scratched the surface of the myriad of ways museums could engage in therapy and social care. Central to this engagement is the enormous value of object handling and touch.

–17–

Measuring the 'Difficult-to-Measure'

Caroline Selai

Introduction

> When you can measure what you are speaking of and express it in numbers you know something about it: when you cannot express it in numbers your knowledge is of a very meagre kind.
>
> Lord Kelvin, *The Physicist*

This chapter introduces some of the conceptual and methodological issues surrounding the rigorous measurement of such nebulous outcomes as the 'benefit' of an intervention – for example, handling museum objects, gazing at a work of art, participating in a therapeutic activity, etc. The chapter will focus on some key practical issues of concern to those wishing to design a study to measure the 'benefit' of activities they are currently carrying out with users/visitors/patients (depending on the local context).

How might researchers set out to understand and measure the value of object handling as an enrichment activity for hospital patients? This chapter introduces the reader to some of the many technical and methodological issues raised. In order to measure the 'value' of object handling, the researcher must first decide what to measure. In the clinical/hospital setting, this might be some aspect of a patient's well-being, health or quality of life (QOL). This chapter concentrates on one of these possible outcome measures, quality of life, and illustrates generic issues important for any outcome measure – which is that a measure must have robust measurement properties such as reliability and validity. When choosing an appropriate outcome measure, researchers must give thought to the most appropriate study design. This chapter give a brief overview of some of the issues involved in designing a study to measure the value of object handling as an enrichment activity for hospital patients.

The Challenges of Measuring the Difficult-to-Measure

Many psychologists and social scientists have spent a great deal of time trying to measure somewhat vague, imprecise phenomena for a number of reasons, in a variety of contexts. In health-care settings, researchers have attempted to define and measure operationally a patient's health status and well-being, mood, stress levels, patient satisfaction and quality of life (to name but a few controversial topics).

The Current Challenge: Methodological Issues

Drawing on this vast literature, how might researchers set out to understand and measure the value of object handling as an enrichment activity for hospital patients? In deciding what to measure, the first question is what beneficial effects object handling might have. These might include improvements in some aspect of patients' overall well-being, health status, mood or quality of life. After having decided what to measure, the next step is to find an appropriate, validated instrument to measure this. Thought needs to be given to the study design and, here, the simplest approach might be to take measurement at baseline, prior to the enrichment activity, and at a follow-up to determine whether the object handling has had any effect, beneficial or otherwise. There are a number of methodological issues. For example, at follow-up, how do we know that any change in well-being is due to the object handling (exclusively) and nothing else? Researchers need to anticipate and control for confounding variables such as age of patient and severity of illness at baseline. I will return to these points later. However, let's go back for a moment to the question of what we might want to measure – perhaps some aspect of well-being or QOL – and how.

This brief chapter will concentrate on the measurement of QOL. This is only one of a number of possible outcome measures but the issues raised, such as the importance of establishing that the instrument has robust measurement properties, are valid for all outcome measures in this context.

Definitions: Quality of Life and Health-related Quality of Life

'Quality of life' has been used in a range of contexts, from the visual arts to conservation and concerns for the environment, from the evaluation of medical treatments to transport and housing policy, from the marketing of products we buy to employment and the quality of working life (Bowling 1995). The earliest population health indices used readily available numerical indicators such as mortality rates (McDowell and Newell 1987). Rising expectations in recent years have led to a shift away from viewing health in terms of survival to a phase of defining it in terms of freedom from disease, to abilities to perform daily activities and, further, to the current emphasis on positive themes of happiness, social and emotional well-being, and quality of life (ibid.). Although QOL has been used across many disciplines (including geography, literature, philosophy, health economics, advertising, health promotion, the medical sciences, sociology and psychology) there is general agreement that, in the clinical context, emphasis should be placed on addressing health-related QOL (HR-QOL). The theoretical framework of health-related QOL is largely based on a multidimensional perspective of health as physical, psychological and social functioning and well-being, derived from the World Health Organization's (WHO) definition of health as a 'state of complete physical, mental and social well-being, and not merely the absence of disease or infirmity'. The WHO set up a working party on quality of life and this group defined QOL as follows:

> Quality of life is defined as an individual's perception of their position in life in the context of the culture and value systems in which they live and in relation to their goals, expectations, standards and concerns. It is a broad ranging concept affected in a complex way by the person's physical health, psychological state, level of independence, social relationships, and their relationships to salient features of their environment. (WHOQOL Group 1993)

Whilst in early chapters, there was much philosophical debate about the definition of QOL and HR-QOL (Fraser et al. 1993; Bowling 1995), it is now only rarely debated in the medical literature (Oliver et al. 1995). However, although there is general agreement about assessing the narrower domain of HR-QOL in the clinical context, and that the three broad domains of physical, psychological and social well-being should be taken as the starting-point of most approaches to HR-QOL assessment, there is a wide diversity of approaches to measurement. The approach taken depends partly on why one wants to measure QOL.

Why Measure QOL?

There may be a number of reasons for wanting to assess QOL and it is important to be clear about these. The type of measurement approach used will depend on the goal of the study and the use to which the data will be put. QOL measures have been devised for a number of reasons (Fitzpatrick et al. 1992).

Applications of quality of life measures (from Fitzpatrick et al. 1992):

- Screening and monitoring for psychosocial problems in individual patient care.
- Population surveys of perceived health problems.
- Medical audit.
- Outcome measures in health services or evaluation research.
- Clinical trials.
- Cost-utility analyses.

Instruments for all of these applications have different approaches to measurement. Before embarking on the development of any new instrument, the investigator should define exactly what the instrument is to measure (Juniper et al. 1996). The investigator needs to decide whether the primary purpose of the instrument is going to be evaluative, discriminative or predictive (ibid.). The type of measurement approach will depend on the type of data required which, in turn, will depend on the use to which the data will be put.

Approaches to Measurement in QOL

There is a fundamental tension in the measurement of QOL. Since what is deemed important for QOL is acknowledged to be subjective and idiosyncratic – differences being influenced by a variety of personal and cultural factors – an appraisal of QOL should strive to capture the individual's subjective, phenomenological experience. On the other hand, the hallmark of scientific measurement is 'objective', reproducible, empirical data collection. QOL researchers have taken a whole range of approaches along this 'subjective/objective' continuum and there now exists over one thousand instruments that have been developed taking a variety of approaches to measurement (Hedrick et al. 1996). The two aspects of the qualitative–quantitative continuum have different strengths. It has been suggested that qualitative methods are more valid whilst quantitative methods are more reliable (Mays and Pope 1996). Approaches to QOL measurement can be considered to fall upon various points of the subjective (qualitative)/objective (quantitative) continuum. They can be split roughly into two groups: (i) the fixed, psychometrically tested questionnaire and (ii) unstructured (or semi-structured), individualized, respondent-driven methods. Of these, the most commonly used is the validated, psychometrically tested questionnaire.

The Fixed, Psychometrically Tested Questionnaire

In the early days, when QOL was introduced into the clinical, or health-care, setting, QOL data were dismissed as being 'soft' data and some clinicians have continued to make this point (Hunt 1997). Partly as a response to this, there was a concerted effort to make QOL measures more 'scientific' and for them to undergo rigorous testing, particularly for use in contexts such as the clinical trial (Juniper et al. 1996). Most effort has been directed at developing the fixed questionnaire and there is a range of QOL/health status measures thus developed (Brooks 1995). *Generic* instruments cover a broad range of QOL domains in a single instrument. Their chief advantage is in facilitating comparisons among different disease groups. Examples include the Sickness Impact Profile (SIP) (Bergner et al. 1981), the Nottingham Health Profile (NHP) (Hunt et al. 1986) and the Medical Outcomes Survey, Short-form-36 (SF-36) (Stewart and Ware 1992; Ware et al. 1993). *Disease-specific* instruments reduce patient burden by including only relevant items for a particular illness, but their main disadvantage is the lack of comparability of results with those from other disease groups.

Health profiles provide separate scores for each of the dimensions of QOL, whereas a *health index*, a type of generic instrument, gives a single summary score, usually from 0 (death) to 1 (perfect health). A further category, developed within the economic tradition, is that of *utility* measures. Whilst some decisions are taken for *individual* patients, others, such as those made by health policy makers, concern *groups* of patients. Here the focus is on *society* as a whole and the societal allocation of scarce resources. For this purpose, preference-weighted measures are required. All of these fixed questionnaires now undergo rigorous testing of their measurement properties. For an example of a generic, Visual Analogue Scale (VAS) see Figure 17.1.

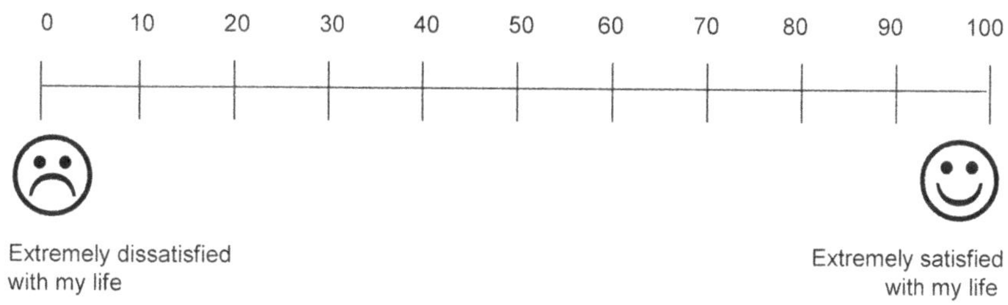

Figure 17.1 Generic Visual Analogue Scale (VAS).

Psychometric Testing of a Questionnaire to Measure QOL or Health Status

The psychometric testing of a questionnaire, for use in a given context, is an intensive, ongoing process (Nunnally and Bernstein 1994; Streiner and Norman 1995). The main issues in instrument development and validation vary slightly depending upon whether the primary purpose of the instrument is going to be evaluative, discriminative or predictive (Juniper et al. 1996). In the health status/QOL literature, it has been generally agreed that a measure should demonstrate at least the basic requirements of validity, reliability and responsiveness, or sensitivity to change (Streiner and Norman 1995).

Validity is how well the instrument measures what it purports to measure. There are various statistical procedures for testing different aspects of an instrument's validity.

1. *Content* validity concerns whether the instrument contains a comprehensive range of items of relevance to the phenomenon it purports to be measuring. In the case of a measure to assess QOL in epilepsy, for example, the question is whether the scale appears to tap items of importance to patients with epilepsy.

2. *Criterion* validity is traditionally defined as the correlation of a scale with some other measure of the trait or disorder under study – ideally, a 'gold standard' which has been used and accepted in the field. Criterion validity is usually divided into two types: *concurrent* validity and *predictive* validity. In the case of concurrent validity, the new scale, or a sub-domain of the scale, is correlated with the criterion measure, both of which are given at the same time. Predictive validity concerns the ability of the test to predict which respondents will achieve a certain outcome at a later date.
3. *Construct* validity is tested by an ongoing process whereby hypothetical 'constructs' or 'mini-theories' are tested. One such hypothesis might be that higher disease severity would correlate with lower scores indicating worse QOL.

Reliability is concerned with whether the same measurement can be obtained on other occasions and concerns the amount of error inherent in any measurement. Two basic tests are the *internal consistency* of a test, measured by coefficient alpha, and *test–retest* reliability where scores taken on two occasions are compared.

There are problems in assessing the test–retest reliability of QOL measures where genuine changes in the patient's well-being may have occurred before the follow-up assessment, making it difficult to distinguish measurement error from genuine change in health/QOL.

Sensitivity is concerned with how sensitive the measure is to detecting small, or clinically relevant, changes. This would be an important property of health status/ QOL measures to be used for monitoring benefits of treatment.

Setting Up a Study

When devising any research project, it is important to be clear about the (tightly defined) research question. It is good practice to set this out in terms of testing a particular hypothesis. In the case of a research project 'to measure the value of object handling as an enrichment activity for hospital patients', this broad aim needs further refinement. There are many issues to consider. Below are some initial thoughts:

- One aspect of the refinement of this research question is to determine what 'value' or benefit there might be of this activity to the hospital patient: for example, an aspect of the patient's health status, mood, quality of life (QOL) or some other aspect of their well-being.
- The research question, already being refined, might now look something like: this study will test the hypothesis that quality of life (QOL) is significantly greater in a group of patients who have received the 'intervention' (object handling as an enrichment activity for hospital patients) compared to this group's baseline QOL

scores and/or to the QOL of patients who did not receive the intervention. This introduces the two additional ideas: (i) recruitment of a 'control' group – a group of patients who did not receive this intervention (i.e. object handling); and (ii) taking measures at baseline and follow-up (thought needs to be given to follow-up interval).

- Next, whatever the chosen indicator of 'benefit' (of object handling), does a robust measurement 'instrument' exist – such as a psychometrically tested questionnaire? Is it valid for the hospital patient group under consideration?
- Thought needs to be given to what other data should be collected – for example, data on the patient's age, gender, diagnosis and co-morbidity, severity of illness, current medication, etc. Researchers need to anticipate and control for confounding variables such as age of patient and severity of illness at baseline. In designing the study, one of the biggest problems is at follow-up – namely, how do we know that any change in QOL, or other aspects of well-being, is due to the object handling (exclusively) and nothing else?
- Thought needs to be given to the recruitment of the patients and, in consultation with a statistician, to the appropriate sample size.
- And so on ...

For an example of a study which has looked at the benefits to hospital patients of object handling using visual analogue scales, see Noble and Chatterjee (this volume).

Conclusions

In devising a study to assess the effects of object handling as an enrichment activity for hospital patients, several aspects need careful consideration. This chapter has focused on thinking about the most suitable 'outcome measure', illustrating some of the key measurement issues with the example of the measurement of quality of life (QOL). Once the researchers have decided what to measure, for example, some aspect of patients' overall well-being or quality of life (QOL), they then need to ascertain whether a validated instrument exists to measure this. Next, the researchers need to give considerable thought to the study design, including method of recruitment of patients and sample size. In a longitudinal design with measurements at baseline and follow-up, what is the most appropriate follow-up interval? What other data need to be collected, for example, in order to control for confounding variables?

This chapter has attempted a brief overview of a large literature spanning many disciplines. Novice researchers are warmly encouraged to consult the existing literature, to seek advice from experienced researchers and to consult a statistician early on. It is always worth spending time carefully planning your study before you begin.

References

Bergner, M., Bobbitt, R.A. and Carter, W.B. (1981), 'The Sickness Impact Profile: Development and Final Revision of a Health Status Measure', *Medical Care* 19: 787–805.

Bowling, A. (1995), *Measuring Disease*, Buckingham: Open University Press.

Brooks, R.G. (1995), *Health Status Measurement: A Perspective on Change*, London: Macmillan.

Duncan, W. (1985), 'Caring or curing: Conflicts of choice', *Journal of the Royal Society of Medicine* 78: 526–35.

Fitzpatrick, R., Fletcher, A.E., Gore, S.M., Jones, D.R., Spiegelhalter, D.J. and Cox, D.R. (1992), 'Quality of life measures in health-care. I: Applications and issues in assessment', *British Medical Journal* 305: 1074–7.

Fraser, S.C.A., Ramirez, A.J., Ebbs, S.R., Fallowfield, L.J., Dobbs, H.J., Richards, M.A., Bates, T. and Baum, M. (1993), 'A daily diary card for quality of life measurement in advanced breast cancer trials', *British Journal of Cancer* 67: 341–6.

Hedrick, S.C., Taeuber, R.C. and Erickson, P. (1996), 'On learning and understanding quality of life: A guide to information sources', in B. Spilker (ed.), *Quality of Life and Pharmacoeconomics in Clinical Trials*, 2nd edn, Philadelphia, PA: Lippincott-Raven.

Hunt, S. (1997), 'The problem of quality of life', *Quality of Life Research* 6(3): 205–12.

Hunt, S.M., McEwan, J. and McKenna, S.P. (1986), *Measuring Health Status*, Beckenham: Croom Helm.

Juniper, E.F., Guyatt, G.H. and Jaeschke, R. (1996), 'How to develop and validate a new health-related quality of life instrument', in B. Spilker (ed.), *Quality of Life and Pharmacoeconomics in Clinical Trials*, 2nd edn, Philadelphia, PA: Lippincott-Raven.

Mays, N. and Pope, C. (1996), 'Rigour and qualitative research', in N. Mays and C. Pope (eds), *Qualitative Research in Healthcare*, London: BMJ Publishing Group.

McDowell, I. and Newell, C. (1987), *Measuring Health: A Guide to Rating Scales and Questionnaires,* Oxford/New York: Oxford University Press.

Nunnally, J.C. and Bernstein, I.H. (1994), *Psychometric Theory*, 3rd edn, New York: McGraw Hill.

Oliver, N., Holloway, F. and Carson, J. (1995), 'Deconstructing quality of life', *Journal of Mental Health* 4: 1–4.

Stewart, A.L. and Ware, J.E. (1992), 'Measuring functioning and well-being: The medical outcomes study approach', Durham, NC: Duke University Press.

Streiner, D.L. and Norman, G.R. (1995), *Health Measurement Scales: A Practical Guide To Their Development and Use*, 2nd edn, Oxford: Oxford Medical Publications.

Ware, J.E., Snow, K.K., Kosinski, M. and Gandek, B. (1993), *SF-36 Health Survey: Manual and Interpretation Guide*, Boston, MA: Health Institute, New England Medical Center.

WHOQOL Group (1993), *Measuring quality of life: The development of the World Health Organization Quality of Life Instrument (WHOQOL)*, Geneva: World Health Organization.

Part V
Knowledge Transfer in Object Handling

–18–

How Accessible Are Museums Today?

Marcus Weisen

This chapter looks at the wider European social and cultural policy context of making objects available for handling in museums to provide, at least, a measure of access to collections for visually impaired people. It presents some of those significant barriers that many disabled people face to the enjoyment of museum collections and experiences of visually impaired people. It identifies a set of conditions which need to be in place (but are not in themselves sufficient) in order to guarantee 'significant and lasting improvements in access to culture for all disabled people', as called for in 1992 by Council of Europe Resolution R(92)6. Irrespective of progress achieved over the last decade, this chapter exemplifies the size of the challenge remaining in order for the cultural rights of disabled people to be guaranteed at last.

This chapter takes a stand against the danger of complacency among decision-making bodies. At a time when 'mainstreaming' has become the buzzword, in processes of organizational modernization we witness the loss of staff in the museums sector with specialisms in social inclusion and access for disabled people.

The preamble opens up a brief discussion about the five senses in literature, art and society. The experience of touch in museums is one of the ways of nurturing a greater diversity of human perception in an increasingly visual world.

Introduction: The Loss of One, the Loss of All

Talk about handling objects often takes place in highly specific professional settings. Conservators handle brittle objects with expert care and handling gloves. Handling collections for schools may have the capacity to arrest the imagination and allow learning by touch. These valuable practices provide the prevailing language for talking about handling. Their specialist focus, however, narrows the scope for talking about the handling of museum objects in any other way. By law of entrainment our language of touch has remained remarkably unimaginative and one-sided.

It is not the object of this chapter to develop a more unbridled and joyful aesthetic, philosophical and anthropological discourse on touch – and its elusive potentials which may forever remain lost to our increasingly ocular-centric societies. However, using a few quotes, I wish to briefly stir, refresh and challenge and show that, in the cultural sector, talk of handling cultural objects needs to connect, at the least, with the hidden depth of thinking around touch that discreetly pervades art and literature.

The French Enlightenment writer, playwright and philosopher Denis Diderot was perhaps the first to give a voice to blind people, quoting them directly rather than talking about them. Asked by Diderot whether he would liked his sight restored, the man replied: 'If it weren't for my curiosity, I'd rather have long arms, for it seems to me that my hands would tell me more about what's doing on the moon than your eyes and telescopes. And your eyes stop looking sooner than my hands stop touching' (Diderot 1749). Was the blind man right or wrong? He was right in claiming the right for his own experience on his own terms. In today's terms, he was asking for a user-centered service. The whole of the cultural heritage sector has not yet woken up to this arresting voice of the eighteenth century, its poetic power and thirst for experience.

Joseph Beuys, one of the twentieth century's most influential artists, said: 'Anyway I'm thinking with my knee'. What a refreshing way to put a Cartesian world view and an overpowering visual culture back in its place. Not only football players think with their knees. Museum and gallery visitors do likewise, as they explore works of art. They protest with their feet when the corridors are infinitely long and only sparse seating is provided – they will simply not come back. The image and memory of the art work that forms in their minds, forms in their bodies too. As a keen walker and a craniosacral therapist, I know that my knees and client's knees have a memory of their own. As we consider the role of touch in museums, we really need to take it out of the ghetto. Today it seems to have become confined to being a kind of 'auxiliary aid' or 'therapeutic tool' for blind people, prisoners and hospital patients, with no relevance for whomsoever else. Great it serves someone, but let's remember those dozens of touch exhibitions for everyone which swept Britain in the eighties and early nineties. As one-off events with no promise of continuity of service they failed visually impaired people, but their philosophy was fundamentally inclusive.

In a poem that is both plain and eloquent, Fernando Pessoa calls for the re-appraisal of the human person as a 'whole body seer': 'I am a herdsman. The herd is my thoughts and my thoughts are all sensations. I think with my eyes and my ears and with my hands and my feet and with my nose and my mouth.' This deceptively simple poem is a powerful programmatic critique of dominant models of cognition. Museums would benefit from rethinking themselves as multisensory spaces.

And what happens in events when the touch taboo, that keeps blind people starved of fulfilling museum experiences, is temporarily lifted? 'I am touching the shoulder and close my eyes: my fingers are unbelievably happy. For the first time in my life they are touching bronze' writes Jean Genet in 'The workshop of Alberto Giacometti' (Genet 1968). So why make touch the preserve of a few groups 'at the risk of exclusion', as if only they needed to be administered it as a special medicine?

The few non-visual and multisensory forms of art, as they have developed over the past few decades to provide access for visually impaired people, contribute to the vitality of art and culture as a whole.

Cultural Equality for Disabled People in Europe: Hitherto a Low Priority

When I talked at the 'Touch and the Value of Object Handling' conference in May 2007, I was Policy Adviser: Inclusion and Communities for the Museums, Libraries and Archives Council (MLA), England. I have written up and expanded that talk as a newly self-employed person. The analysis remains largely the same. I now have the pleasure of expressing opinions free of any role restrictions.

Access to cultural heritage for disabled people is today being discussed everywhere in Europe, from Portugal to Russia and Greece to Norway. Anti-discrimination legislation is making an appearance, and in some countries there exists a right or limited rights of access to services for disabled people – most notably in the UK. Cultural policies designed to provide equal access for disabled people are beginning to appear in several countries. But how accessible are cultural organizations today, after some twenty-five years of pioneering good practice and emerging policy initiatives?

The uncomfortable truth is that, in Europe as a whole, very little evidence has been collected to document, assess and understand the degree to which disabled people face barriers to cultural life and cultural exclusion.

This fact is in itself symptomatic of the low – and, in any case, insufficient – priority given to the cultural equality of disabled people by most, if not all, national governments. In 2003, the European Council agreed the 'Council Resolution on accessibility of infrastructure and cultural activities for people with disabilities'. The Resolution included a commitment to monitor progress achieved so far in 2005. In 2008, work on this progress report is not yet started. This is but one of many examples of the limited degree to which cultural equality of disabled people has become an integral part of the functioning of cultural institutions and organizations.

Meanwhile Disabled People Keep Facing Barriers

Compelling evidence about barriers and discrimination abounds (even if in a form not recognized by law as being discrimination), if we care to listen to the experiences of disabled people:

'I will never go back to a museum. There is nothing about my culture. It is as if deaf people did not exist.'

'The theatre is new and physically accessible, but staff behave as if I came from another planet. I prefer to go to the other theatre, even though staff need to carry me up a few stairs.'
'The person at reception was hopeless! They didn't have any idea about what services they had available for visually impaired people.'

Quotes like these reveal a truly sad state of affairs. Let's also look at literary quotes of self-experience. The following one is by Jorge Louis Borges, Literature Nobel Prize Winner and third visually impaired Director of the Argentine National Library:

> I always thought of paradise as a library. Others think of it as a garden, or perhaps a palace. And here I was, so to speak, in the middle of nine hundred thousand books in diverse languages. I realized I could hardly make out the titles and the spines ... we were two learned men walking through this Library of forbidden books. (Borges 1985)

Borges's metaphor is one of being expelled from Paradise. And indeed, cultural equality of disabled people is like a Paradise that most disabled people have never had. It needs to be gained or re-gained. Everyone who is committed to cultural access and equality for disabled people will agree that the illegible titles on the spines in libraries and dimly lit tiny labels in museums can be avoided, and that practical and aesthetically pleasing ways exist to do this.

Compelling Statistics about Barriers to Cultural Heritage

Strategic cultural organizations in Europe have barely begun to collect evidence of the experiences of disabled people, the barriers they face and the practical guidance they can contribute to improvement. The majority of such research and surveys has been undertaken over the past twenty years by disability organizations – a reflection on wittingly or unwittingly discriminatory practice in cultural research. Here are just a few salient findings:

1. In 1988, 1 per cent of museums provided some (and usually by no means a sufficient proportion of) information in Braille (Museums Association Annual database (now defunct) and Royal National Institute of Blind People, UK. Sample: 1,300). In 1998, 4 per cent did (Museums and Galleries Commission (now MLA) Domus survey, England. Sample: 900). Can we call this progress? Sadly, the UK is far from an isolated case. On the contrary, it is recognized by many as one of the more advanced countries in the area of access to culture.
2. Less than a third of reception desks had a loop system (Scope UK 2000 – working for people with cerebral palsy). People who are hard of hearing are by far the largest group of disabled people. Hearing loops are very affordable (GBP105–155; Euro 150–225). And yet, the cultural sector remains largely unaware of their requirements.

3. Only about 2 per cent of museum websites in England meet the web accessibility standards required by European e-government policies. The international standard is called Web Content Accessibility Guidelines (WCAG). The research had been undertaken by City University and commissioned by the Museums and Galleries Commission (MLA) in 2005. If we want to gain an understanding about barriers to cultural websites, we need to be aware that the WCAG do not cover the requirements of, for example, deaf people. The sad story is that only a very tiny proportion of cultural sector websites provide basic visitor information in Sign Language to deaf people. The presentation and interpretation of collections to specific groups of disabled people is not covered either by WCAG. Here, the barriers are compelling: only a tiny fraction of cultural sector websites provide, for example, verbal descriptions of online objects aimed at visually impaired people.

The examples given here reveal the tip of the iceberg, for barrier-free accessibility is always a complex and multidimensional reality.

Barriers Are a Multidimensional Reality

- Barriers can be: attitudinal, institutional, physical, sensory, intellectual.
- Barriers can be economic, about lack of choice, lack of accessible transport or due to a lack of educational opportunity.
- Barriers are interdependent.
- Barriers can be fluid in a fast-moving world – for example, as the UK gets something of a handle on physical barriers in cultural venues, it faces the new global challenge of web accessibility.
- Repeated experience of unnecessary barriers leads to frustration, anger, resignation and finally cultural exclusion.

Usually only a selection of objects, exhibitions and spaces are, to a greater or lesser extent, barrier-free to disabled people. As an example, the information that is accessible to visually impaired people only represents a fraction of a fraction of a fraction of all the information provided by museums. As a conclusion, the creation of barrier-free spaces and services requires a systematic approach. Once again the truth is uncomfortable – such an approach is lacking in most cultural venues in Europe.

The Experiences of Visually Impaired People

The following user quotes highlight the vital importance of staff training and organizational procedures in the service of access. They are taken from one of the few research reports about the experiences and desires of visually impaired people (RNIB/Vocaleyes 2003):

'At reception people haven't said anything. I didn't know people were there and they would probably be expecting someone to make eye contact with them.'
'When an organization has produced an accessible audio guide, they should tell their staff.'

Non-users of museums, galleries and heritage sites gave a range of reasons for not visiting or attending:

- Twenty-five per cent said they were not interested;
- Twenty-four per cent said there would be little to enjoy for visually impaired people;
- Eighteen per cent said it would be difficult to find transport to the venue;
- Fifteen per cent said it would be difficult to find a companion.

A few simple observations can be made:

- Few museums publicize accurately what kind of services they offer. (Guidelines accompanying the research report provide recommendations that address this issue (RNIB/Vocaleyes 2003).
- Museums which 'crack' the transport issue, are likely to be successful in widening visually impaired audiences.
- Marketing of services should also be addressed at sighted companions.

With regard to the question of what would make museums, galleries and heritage sites more attractive, respondents suggested:

- Wider availability of specialist facilities (84 per cent);
- More general information (83 per cent);
- Improved quality of facilities or events on offer for visually impaired people (82 per cent).

The advice is clear and good programmes for improvement could be devised based on these comments. When questioned about what information and services did visually impaired people find very helpful? The most quoted services were:

- Displays and objects that can be touched;
- Models that can be touched;
- Audio guides with descriptions for visually impaired people;
- Live events or talks;
- Raised images;
- Tactile maps.

This chapter focuses on raising awareness of barriers because of our innate ability to skim over them, but I will end this section with the kind of happy quotes that reward a museum job well done. The quote comes from Peter Bosher, a Partner of Insight Training for Museums, Galleries and Heritage:

> The audio guide made all the difference in bringing this exhibition to life. I've never spent so long viewing a single exhibition and when I say 'viewing' as a totally blind person that means something rather extraordinary. I actually felt I'd experienced the Holocaust Galleries directly, which is rare if not a first. This was largely due to the clear, relevant, well presented commentaries and descriptions in the guide. (personal communication)

A Compelling Ethical Case and International Policy Statements

Access and inclusion are indicators of how civilized our European societies are. Are we happy with what has been achieved so far? Are we civilized enough? Is Europe civilized enough? Let us also seek inspiration from Mahatma Ghandi who, in a speech in South Africa in 1890, said:

> A customer is the most important visitor to our premises. He is not dependent on us. We are dependent on him. He is not an interruption on our work. He is the purpose of it. He is not an outsider on our business. He is part of it. We are not doing him a favour by serving him. He is doing us a favour by giving us opportunity to do so.

Unfortunately disabled people are still frequently seen as a nuisance and an impediment to the 'normal' functioning of cultural organizations. Directors and managers would of course emphatically deny this, but, as the old proverb says, 'the proof of the pudding lies in the eating' and in our case, the pudding would show as barrier-free and inclusive services.

The ethical argument has found its way into the Universal Declaration of Human Rights (1948, article 27.1): 'Everyone has the right freely to enjoy the arts and the cultural life of the community', and, much later, into the UN Standard Rules on Equal Opportunities for People with Disabilities (1994, article 10) 'States will ensure that persons with disabilities are integrated into and can participate in cultural policies on an equal basis'.

The little known Council of Europe Recommendation R(92)6 of 1992 is of particular importance because it recognizes the necessity for systemic and systematic change: 'Government institutions, leisure and cultural organizations should develop comprehensive access policies and action programmes designed to significant and lasting improvements for all people with disabilities.'

UK Policy and Dynamics for Change

In spite of all the evidence of barriers highlighted in UK research, the UK is seen as a leading country in access for disabled people to culture. For example, no other European country has broadcasting legislation which recognizes the rights of visually impaired people. In the UK, on the other hand, there are hundreds of hours of audio-described television programmes each month. Yet, this dimension of cultural life has barely reached museum websites, with only a handful providing descriptions of online presentation – thus showing that the cultural offer is not user-centred and there is no notion of seamless services.

Two policy developments in the UK have powerfully influenced the trend towards the cultural equality of disabled people (MLA Disability Survey 2005). Firstly, the Disability Discrimination Act 1995, which gives disabled people a right of access to services and makes refusal of service unlawful. The UK is also one of the few countries in which cultural policy is influenced by social policy. Cultural organizations have been expected since 2001 by the Department of Culture Media and Sport to be 'agents of social change', to combat social exclusion and provide equal opportunities.

Government policy often echoes and formalizes grass-roots developments and the creative engagement of government agencies. Thus a real turning point was when, in the early 1990s, the Arts Lottery made access for disabled people a criterion for cultural funding. This has been achieved thanks to the stubbornness of disability activists and was by no means a foregone conclusion. It was a vital and necessary step in the inclusive budgeting direction, a principle which can hardly be said to have been successfully 'mainstreamed'. In 2001, 19 per cent of respondent museums, libraries and archives allocated revenue funding to disability access (MLA Disability Survey 2002), with 30 per cent doing so in 2005 (MLA Disability Survey 2005). Without inclusive budgeting, any statements of commitment to the cultural equality of disabled people remain empty words.

Aspects of the cultural sector's professional values and cultures are a dynamic factor. These specific professions have a service tradition that is generally welcoming to disabled people, such as museum educators and theatre house managers. The cultural sector has also published a range of good practice guidelines – although gaps remain and up-dating continues to be an issue.

Factors such as these are a requirement for lasting and significant changes to come about, as called for by the Council of Europe, but they do not provide a guarantee that significant and lasting changes occur.

Disability Discrimination Act: Linking Organizational Good Practice with an Anticipatory Duty

Research has shown that cultural organizations which take a planned approach to change are much more likely to be more accessible to disabled people. In an MLA Disability Survey (2002), nine out of ten 'high performers' (scoring 85 or more out of 115 indicators) had carried out an access audit and provided disability training, compared with less than a third of 'low performers'. They were also ten times more likely to consult users and non-users.

Of the respondent museums, libraries and archives, only 27 per cent stated that they had an access plan in 2001 (MLA Disability Survey 2002), whereas 43 per cent of respondents of the 2005 survey had one (MLA Disability Survey 2005). In 2001, 67 per cent respondents had undertaken an access audit, while in 2005 the figure had risen to 84 per cent.

The DDA puts an 'anticipatory duty' on service providers to know what access requirements disabled users may have. The related Code of Practice gives disability access policies, plans, audits and training, as well as consultation with disabled people, to demonstrate how service providers may meet this duty.

The correlation between these practices and effective performance in access for disabled people shows how judiciously chosen this advice from the Disability Rights Commission is.

Conclusion: The Way Forward by Systematic Change

Access for disabled people can only be realized successfully if it becomes integral to everything a museum does. As we begin to get a handle on existing barriers (e.g. through environmental access), new barriers are appearing such as those to virtual access. Access is an ongoing process and requires a systematic approach. Whilst we can celebrate the many recent changes, we are under no illusion that there is still a long way to go. What the outgoing minister for culture, David Lammy, has written about the cultural sector holds true for society as a whole: 'There are many challenges remaining and we must maintain a sense of urgency as we continue to improve the quality of life of people with disabilities through enabling their fundamental right to have access to cultural activity' (*Times*, 7 June 2007).

References

Borges J.L. (1985), 'Cécité' (Blindess), in: *Conférences*, Coll. Folio Essais, 1985, Paris: Gallimard.

Diderot D. (1749), Pensées Philosophiques, Lettre sur les Aveugles, Supplément au Voyage de Bougainville, Paris: Garnier-Flammarion.

Genet J. (1968), 'L'atelier d'Alberto Giacometti' (Alberto Giacometti's workshop), L'Arbalète.

Museums, Libraries and Archives Council/City University (2005). Accessibility of museum, library and archive websites: the MLA audit.

Museums, Libraries and Archives Council (2005). *Web Accessibility Survey.* www.mla.gov.uk/website/policy/Diversity/Web_Accessibility, accessed 1 February 2008.

Museums, Libraries and Archives Council (2002). *Disability Survey.* http://www.mla.gov.uk/website/policy/Diversity/People_With_Disabilities/, accessed 1 February 2008.

Museums, Libraries and Archives Council (2005). *Disability Survey.* http://www.mla.gov.uk/website/policy/Diversity/People_With_Disabilities/, accessed 1 February 2008.

Pessoa, F. (1911–1912) *Le Gardien de Troupeaux* (The Shepherd), Poésie, Paris: Gallimard.

Royal National Institute of Blind People (RNIB) (2003), 'Talking Images Research. Museums, Galleries and Heritage Sites: Improving access for blind and partially sighted people', www.rnib.org.uk; search on 'talking images', accessed 1 February 2008.

–19–

The British Museum in Pentonville Prison: Dismantling Barriers through Touch and Handling

Jane Samuels

Introduction

This chapter provides a brief overview of a recent partnership between the British Museum, a museum of world culture based in Bloomsbury, London, and Pentonville Prison, a male prison located in North London. This inspiring partnership took place between 2005 and 2007 yielding impressive and powerful results.

The British Museum has as one of its key objectives the aim of reaching the widest possible audience. Consequently the development of partnerships with under-represented audiences is commonplace. This partnership was conceived in 2005 by the Access Manager in the Learning and Audiences Department at the British Museum and the Head of the Learning and Skills Department of Pentonville Prison. The aim was to draw from the vast collections of the British Museum to forge a creative and multisensory project which would be intellectually accessible and engage prisoners on many levels – socially, creatively, intellectually and culturally.

A key concern was how to engage the prisoners and sustain their enthusiastic involvement throughout the life of the project. A variety of contrasting museum objects and exhibits were chosen to enliven the project with the hope of attracting and sustaining the interest of the prisoners. Further, the provision of technical equipment so that the prisoners could film their work and create their own audio tracks became another form of handling and thus an integral part of the process of engagement.

The project also selected subjects which would interest the prisoners and yield powerful responses. Through the 'Throne of Weapons', a chair made from AK47s, the prisoners explored and discussed gun crime, conflict, peace and resolution; and through the second exhibit, 'Cradle to Grave', an installation constructed primarily from pharmaceutical drugs, the prisoners explored drug use, well-being and health.

The prison's chapel was selected as the location for both the exhibitions and projects. The reason for choosing the chapel was not a religious one, but was motivated by its large, quiet space which, as the venue for Sunday services for both Roman Catholic and Church of England prisoners, attracted a substantial weekly audience. Maximizing exposure to both exhibits was important to the project objectives; thus, by placing them in a widely frequented location, approximately 500 prisoners a week were exposed to the 'Throne of Weapons' and "Cradle to Grave" exhibits. At full capacity the prison holds approximately 1,200 men.

To increase exposure further, the prison's Learning and Skills Department worked with other faith communities, Sikh, Muslim, Jewish, to ensure they also had the opportunity to experience both exhibitions and did not feel excluded by the exhibits location.

Phase 1: The 'Throne of Weapons' in Pentonville Prison

The 'Throne of Weapons' (Figure 19.1) was the first exhibit to be installed in the chapel of the prison in September 2005. The 'Throne of Weapons' is a piece of

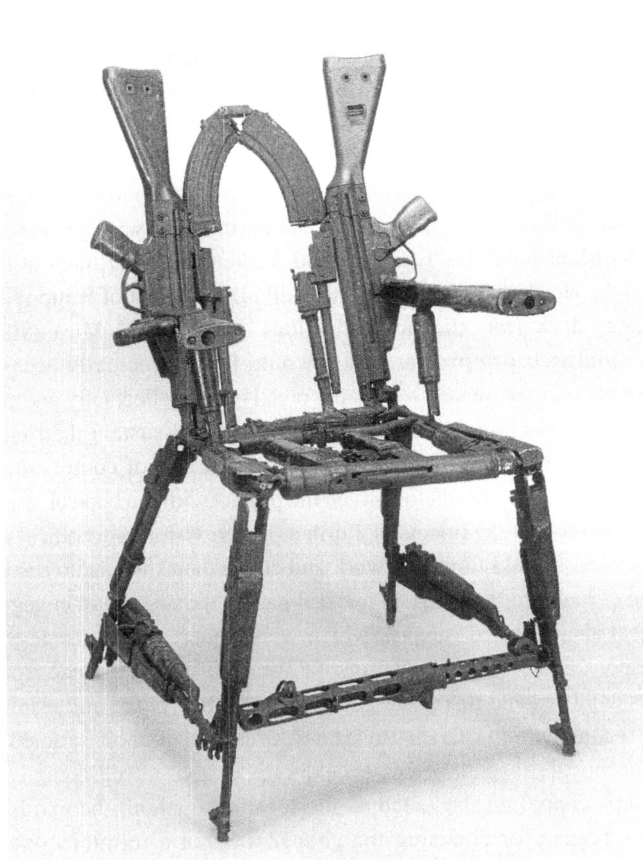

Figure 19.1 The Throne of Weapons © Kester. Image © The Trustees of the British Museum.

Mozambique contemporary art. It was created by Mozambique artist, Kesta, using recycled decommissioned weapons handed in during a guns amnesty following the country's sixteen-year-long Civil War. The aim was to transform 'weapons of war into icons of peace' (Christian Aid 2002). These weapons were handed in following the civil war in exchange for practical objects such as tools, ploughs, fertilizer and sewing machines. This initiative, known as 'Transforming Arms into Tools' was set up by Christian Aid in an attempt to promote peace and help rebuild a war-ravaged country. The dismantled guns which form the 'Throne of Weapons' were handed in so that shattered lives could be rebuilt.

> Weapons are handed in, no questions asked in exchange for productive tools ... Some of the most commonly received weapons are AK47s, Mausers, grenade launchers, pistols, bazookas and semi-automatic rifles. (Christian Aid 2002)

The 'Throne' commemorates the action of giving up something inherently destructive in exchange for objects or materials with constructive and positive value. It is therefore interpreted by many as a symbol of peace and reconciliation.

The exhibiting of the 'Throne' in the Chapel of Pentonville was accompanied by a film-making project facilitated by two practising artists, a curator from the African collections and coordinated by the museum's Access Manager. The project group was made up of fourteen prisoners representing a range of ages and ethnicities – English, Irish, African and Afro-Caribbean (John Holden 2005). Most of the participants were on remand for a wide range of offences, from the relatively minor to the extremely serious. A few had been convicted of their crimes and were awaiting deployment into other prisons.

Many of these prisoners more commonly spent up to twenty-three hours a day in their prison cells. Therefore, the opportunity to express their thoughts, hold and touch objects from the collections of the British Museum and experiment with a variety of creative mediums was embraced by most with unreserved enthusiasm. The prisoners discussed their experiences of, and feelings about, gun use. Many of the men revealed they had been nurtured on a diet of gun crime from a very early age. In response to the themes discussed, the men wrote poetry and prose, created visual art (Figure 19.2), danced, sang and composed their own music. Some of these men were talented, producing outstanding work of an unexpected quality.

Nothing could have prepared the project team for the unforgettable and impressive creations which flowed so freely from the prisoners. John Holden, who was evaluating the 'Throne of Weapons' UK Tour as part of a programme called Africa 05 commented:

> [T]he group's creativity was abundantly clear. Not only were they motivated, they were talented, doing great work with video and the written word, working towards a final product that would be emotionally moving and visually striking. (Holden 2005)

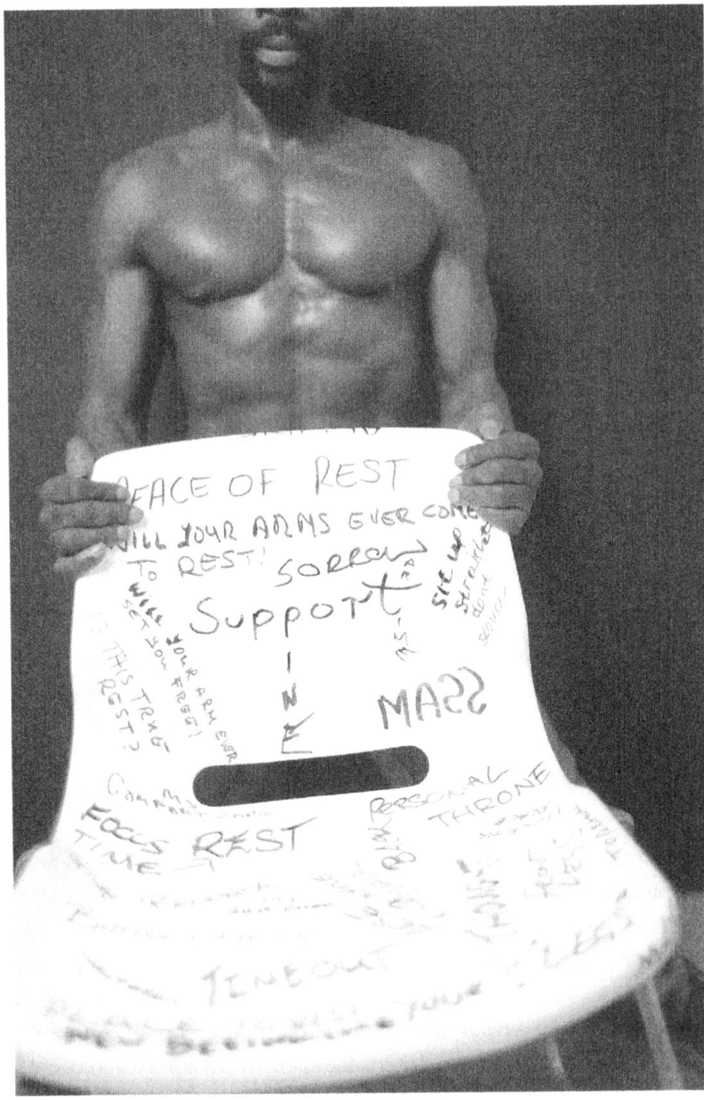

Figure 19.2 Inspired by the Throne of Weapons, chair inscribed with the words of the prisoners. Photograph by Benedict Johnson, 2005. © The British Museum.

Phase 2: 'Cradle to Grave' in Pentonville Prison

Equally impressive results evolved from the second phase of the project which took place the following year in September 2006 and centred around a reinterpretation of 'Cradle to Grave', an installation originally created for the museum's Wellcome

Trust Gallery of Living and Dying. This powerful installation was conceived and created by Pharmacopeia, a partnership of three artists – a textile artist, a film-maker and a general practitioner – whose knowledge of medicine helped form an accurate profile of medicinal use for the main content of the installation. This smaller sister version of 'Cradle to Grave' to inhabit Pentonville Prison was built specifically for the project and told the story of one man, named Mr A.N. Other, and his experience of illness and well-being over the course of twenty-eight years. Once again, fourteen prisoners formed the participant group.

The installation measured 5 metres long by 130 centimetres and featured Mr A.N. Other's intake of 3,120 pills over twenty-eight years, including, Junior Paracetamol for pain, Amoxicillin for infection, Prednisolone for asthma, Oxytetracycline for acne, Doxycycline for malaria, Viagra for impotency, Prosac for depression, and Efavirenz AR and Combivir AR antiretroviral drugs (Figure 19.3).

The installation also featured a variety of photographs, personal possessions and objects which represented Mr A.N. Other's life over the course of the twenty-eight years, including a hearing aid, a pair of glasses, a glass of red wine, a Bible, an asthma inhaler and an ash tray filled with cigarette butts from the security guards at the prison.

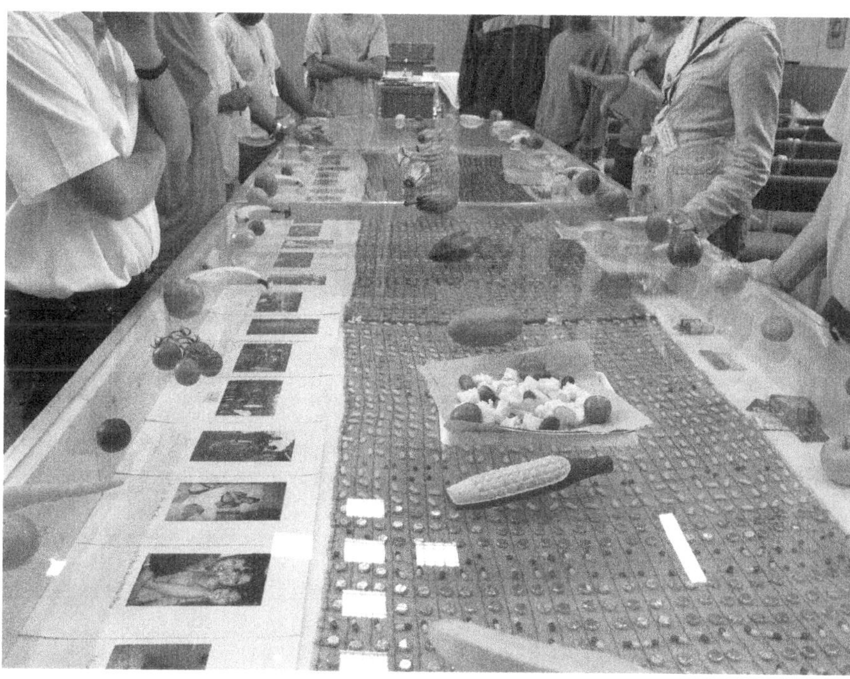

Figure 19.3 A reinterpretation of the original Cradle to Grave in the Chapel of Pentonville Prison. Photograph by Susie Freeman, 2006. © The Trustees of the British Museum.

During his lifetime, Mr A.N. Other experienced various conditions, including asthma, hay fever, sports injuries, malaria, and the contraction of HIV at twenty-seven. Here, his life story ends and the audience is left wondering whether contracting HIV resulted in AIDS and thus his premature death.

For the fourteen participating prisoners, the installation helped break down barriers and provoked a great many responses. Prisoners would point at familiar pills and exclaim, 'I take that medicine. What is it?' It also prompted them to discuss their own life experiences, revealing when they lost their virginity, or at what age they experienced their first significant death of a close relative. Many of the experiences of the prisoners were unremarkable, others however were so heart-wrenching that it was unsurprising their lives had led them to prison.

The project concluded with an exhibition and performance by the prisoners using poetry, visual art, film and music. The most striking contributions were often powerful and emotive poetry written by many of the men. In the confines of their cells, the written word was an important creative outlet which remained possible; many of these men would therefore go back to their rooms at the end of a workshop and return the following day with a newly penned poem to contribute to the project. These poems often revealed the prisoners hidden vulnerabilities or grotesque experiences of abuse. The following poem expresses one prisoner's feelings about imprisonment:

Banged Up
Banged up behind doors, treated like an animal
With all rights denied
And feeling misery from inside
As from window cells I can see the outside
But I can't embrace.

Banged up with numerous anxieties
Imagining things I used to take for granted
All of which I've lost
And I can't embrace.

Banged up with tension and surrounded with rudeness,
And with the madness of the environment I'm in
Which I can't fight
And I can't embrace.

Banged up with frustrations
Thinking of the day
On which I'll be released
For the day is delayed
When I can embrace.

Banged up with sadness
Of the way I'm ignored
And the way I'm told
The things I've got to do
And when I've got to do them.
But I learn to remain calm
For the time of release
Is near to embrace.

Author anonymous

The Power of Touch

A key component of both projects in Pentonville was the use of touch and handling. This came in three forms. First, the prisoners were encouraged to interact with the main exhibits, the 'Throne of Weapons' and 'Cradle to Grave'. Second, they were invited to hold and touch a variety of additional artefacts brought into the prison by a curator from the museum's African collections. Finally, the men were given the use of digital cameras, art equipment and computers to facilitate the development of their creative ideas. The freedom to touch and hold a variety of contrasting artefacts from the museum's collections, and the use of a range of creative and technical equipment to help the project participants realize their artistic objectives, was critical in engaging the prisoners and sustaining their involvement.

This multisensory dimension to the project was key to achieving its objectives. Entrusting the prisoners with valuable objects and equipment made them feel respected, and thus the facilitators in turn gained the trust and acceptance of the prisoners – a crucial requirement if both projects were to be successful. Jenny Thomas, a leading educator from the Learning and Skills Department at Pentonville, was present throughout both projects. She was surprised by how cooperative and relaxed the men were:

> When the men initially volunteered to attend the workshops I was concerned at the mix of personalities and their commitment to attend each day of the project. As the workshops progressed, it was clear to see they had engaged thoroughly and enthusiastically with the issues surrounding the exhibits. Any personal differences or problems were put aside in pursuit of their common goal. Each man actively and creatively responded to the Throne and 'Cradle to Grave': some working alone – some in collaboration. Their attendances at the workshops were uniquely high. (Jenny Thomas, Teacher, Department of Learning and Skills, Pentonville Prison)

The project team observed how the men thrived and prospered from this rare and unique learning opportunity. One enthusiastic inmate insisted that his time on the project had been a 'life changing experience'. Another said the project was the first time he had felt 'like a human being [since arriving in Pentonville]'. Very encouragingly, some of the men did initiate visits to the museum following their release from prison. This unexpected outcome was a powerful indication of the positive impact of both projects on at least some of the prisoners.

Conclusion

The use of objects and exhibits had a beneficial impact on the morale of the prisoners, helping to dismantle barriers between everyone – prisoners and staff alike. It encouraged the prisoners to trust the project team, enlivened the learning experience and enhanced the creative process and final outcome.

The participants, who in many instances live harsh and unimaginable existences, softened their usual tough exteriors to contribute to the creative process with carefully considered poetry and music and delicately crafted paintings and drawings. Touch is a powerful medium, and in this context the use of objects and other equipment helped the prisoners to open up and communicate with one another. In the usual culture of prison life, inmates would be hesitant to drop their guard so willingly, but in this unique project environment, away from the prying eyes of the rest of the prison and with the help of objects to excite and illuminate, the prisoners put down their barriers and became, if only for a limited time, one harmonious team working towards a common creative goal.

The positive impact of the objects and exhibits on the prisoners was most strikingly expressed by one prisoner who was on remand for murder. Moved by the 'Throne of Weapons', he told other members of the group 'the most powerful thing you can do is pick up a book not a gun' (Holden 2005: 26).

References

Christian Aid (2002), 'Swords into Ploughshares: Transforming Arms into Art', Exhibition brochure, London: Christian Aid.

Holden, J. (2005). '"Throne of Weapons", A British Museum Tour', London: British Museum, A Partnership UK Project.

–20–

The Amenable Object: Working with Diaspora Communities through a Psychoanalysis of Touch

Bernadette Lynch

Introduction

> I want you to touch me on my inside part and call me my name.
>
> Toni Morrison, *Beloved*

Using collections to build ethical and sustainable relationships with people is increasingly accepted as an important social role for museums today. This chapter argues that it is in this relationship work, which takes time, commitment and sustained effort, that 'touch' comes into its own. Yet, despite the primacy of touch for many cultures, the ethos of the Western museum – centred as it is on vision alone – negates sensory experience and, in particular, the tactile sense.

From the philosophical and emotional to the practical applications of touch, this chapter proposes a new understanding of the museum object. Drawing on psychoanalysis, the importance of 'active' viewing and sensory experiencing of the ambiguous 'amenable object' in the museum is emphasized as a way of powerfully communicating between people from very different worlds.

Thus the chapter proposes a wider, more in-depth use-value for the museum object, with recommendations for the implementation of a new sensory museology. The suggestion is made that museums have much to learn from the exploration of a new practice based on the senses, with the potential to change both museums and their future relations with audiences.

What is the relationship between the image of the world provided by the Western empirical sciences (as is deeply embedded in museum practice in the West), and the way in which the world is actually experienced in everyday life, all over the world? In other words, how can museums bridge the separation between the 'scientific' and the 'lived' worlds, particularly at a time when, increasingly, the 'stranger', as in diaspora communities, is 'entering the house', as Homi Bhabha (1994) put it, and disrupting precedent, particularly through challenging the touch taboo? Is it possible this stranger can teach the museum an entirely new appreciation for the use-value of museum objects?

But museums may, as yet, be unaware of the ethical importance and use of *touch* in developing and maintaining relations with diaspora communities.

In order to begin to appreciate the potential for touch in the museum, it is necessary to examine something very complex – how people 'feel' when they are in museums, in every sense of the word, and how the museum can facilitate this experience through direct access to collections. 'Direct access' in this case means 'object handling' and its central importance, it is maintained here, in developing a culture of engagement with museum collections.

The work of phenomenologists such as Merleau Ponty (1962) emphasized *active engagement in the world* by individuals exercising their capacity to make choices and impose meanings. It requires a radical questioning; it demands that we question what we see before us and become deeply suspicious of vision alone, and thus has implications for the singular reliance on vision in museums, typical of the museum visit as we glide by or stand transfixed before the vitrine.

For the phenomenologist Merleau Ponty, there is nothing more difficult to know than what we see before us, distorted as it is by the 'prejudice' of objective thought. In contrast, he proposed the 'lived world' *as experienced.* In museums this might involve entering into an attitude of methodic unknowing where things cease to be facts, data, objects, possessions, and become *questions* ... 'simultaneously remind[ing] us that all radical questioning takes its point of departure in our ... experience of the life-world' (Kearney 1991: 5).

Being in Diaspora

Being confronted with the data, facts, objects in a Western museum can intimidate many people, so what is the experience for those diaspora communities for whom the entire cultural code of the museum may be utterly unfamiliar, intimidating and totally inaccessible? First, let us first look at the experience of being 'in diaspora' which can be defined as:

- Any group migration or flight from a country or region; dispersion;
- Any group that has been dispersed outside its traditional homeland;
- Any religious group living as a minority among people of the prevailing religion.

Ian Chambers describes the limbo of the here and not here experience of diaspora:

> [L]iving between worlds, caught on a frontier that runs through your tongue, religion, music, dress, appearance and life. To come from elsewhere, from 'there' and not 'here', and hence to be simultaneously 'inside' and 'outside' the situation at hand, is to live

> at the intersections of histories and memories ... It is simultaneously to encounter the languages of powerlessness. This drama, rarely freely chosen, is ... the drama of the stranger. (Chambers 1994: 6)

To be in diaspora does not mean to be in a place which has simply transplanted the original home onto foreign soil. As one commentator put it, 'Diasporic culture cannot be seen in organicist terms, as if it were a seed that could be transported and planted elsewhere' (Papastergiadis 1998: x1). In the aftermath of colonialism, migrancy and exile have become potent motifs of modern culture and 'we find ourselves in the moment of transit where space and time cross to produce complex figures of difference and identity, past and present, inside and outside, inclusion and exclusion' (Bhabha 1991: 1). Edward Said writes:

> The exile knows that in a secular and contingent world, homes are always provisional. Borders and barriers which enclose us within the safety of family territory can also become prisons, and are often defended beyond reason or necessity. Exiles cross borders, break barriers of thought and experience. (Said 1990: 365)

Refugees, in particular, can experience this as a visceral fear – the eyes lowered, shoulders stooped, in an attempt at invisibility, while all the time under pressure to adapt in order to survive. The writer Vikram Seth describes the 'separateness and fear' attached to the self-conscious acquisition of English (Seth 1994: 65). To speak in the way desired is to begin the process of speaking against oneself, conceding that one's 'tongue is warped' (ibid: 68) in a form of derivative existence. Isolated within this strange semi-life, the exile misses many things. Indeed:

> the loss of familiar signs makes every turn problematic. Disconnected from the place where action was guided by experience, the exile is catapulted into social space where nothing can be taken for granted ... It is narcissism's energy and the momentum drawn from the recognition by others that is undermined by exile. (Papastergiadis 1998: 4)

The borders shift all the time, between generations, between men and women, between cultures.

> Migration is a process which is not completed by the arrival of an individual in a foreign place. Arrival rarely means assimilation. Migrants are often transformed by their journey, and their presence is a catalyst to new transformations in the spaces they enter. Similarly their relationship to their original 'homes' is rarely erased. (Papastergiadis 2000: 205)

At the same time, as James Clifford puts it, 'diaspora cultures [frequently] work to maintain community, selectively preserving and recovering traditions, "customising" and "versioning" them in novel, hybrid, and often antagonistic situations' (Clifford 1997: 263).

Arriving, as they may, at the doors of the Western museums, it would be a mistake for the museum to see these communities as somehow 'fixed' in their separate identities – be it Somali, Kurdish, Afghani or Pole. Said and Bhabha warn of the dangers of fixing and fetishizing identities and the 'calcification of colonial cultures' with a fixation on 'roots' and the romance of the past, or by creating an idea of a homogenized present.[1] Bhabha speaks of the importance of the process of 'presencing' beginning in diaspora, when, as he puts it, the 'unhomeliness' of the present situation is captured, which is a 'condition of extra-territorial and cross-cultural initiations' (Bhabha 1994: 9).

What Has the Museum Got to Offer?

Where museums may have a role is in recognizing what Bhabha describes as 'the borderline work of culture'. Diaspora communities can help museums, and in the exchange negotiated within this borderland, be helped by the museum, so that, as Clifford put it, 'master narratives of cultural disappearance and salvage [can] be replaced by stories of revival, remembrance and struggle' (Clifford 1997: 109).

It is in this area, this borderland, where relationship work with individuals in diaspora communities is negotiated, taking sensitivity, time, commitment and sustained effort of all staff, that, I will argue, 'touch' comes into its own.

The museum can learn from the challenges presented by diaspora communities who have not learned the museum's prohibitive codes. It is through the sensory exploration of objects – of 'object relations' (Klein 1932) in the museum, in the sense of internal and external 'objects' – that the individual in diaspora can exercise their need to remember, repeat and work through their feelings of depression and anxiety. As the Sandlers remind us:

> The urge to re-experience important subjective aspects of object relationships from the first years of life constantly recurs and persists (i.e. represents unconscious wishes) particularly when our feelings of security or safety are threatened. (Sandler and Sandler 1978/1999: 285–96)

Thus, for example, a group of Somali refugee women arrive at a museum (as happened at Manchester Museum in 2000) in just such a state of anxiety and insecurity, and tentatively begin telling their stories, as they attempt to 'work through', negotiate and communicate new relationships in a strange land.

It can be understood as their search for a suitable story to tell themselves and others, for we must remember that in cases where people have undergone terrifyingly dramatic, life-altering situations, the stories told to themselves no longer even hold an appearance of continuity, the illusion of life's 'progress' to which most of us cling. 'Regular life' is thus clearly severed, blown apart. In these cases, the urge to recount the story can be very great indeed, significantly adding to the responsibility of the museum to be extremely careful in its ethical handling of the psychological impact of such experiences, while also appreciating the contingent nature of these narratives.

This is *not* a process of excavating the women's stories, uncovering 'hidden histories', as though 'lost from view'. These shifting 'narratives-of-self' in diaspora can only be constructed through the availability of an audience willing to hear and to understand them, as, at that moment, they intend they should. 'This aspect of the remaking of a self in the aftermath of trauma highlights the dependency of the self on others' (Brison 1999: 46), thus clearly demonstrating the situation of dependency upon others described by the object relations theorist, Melanie Klein, that all of us experience to varying degrees in our lives (Klein 1932). It explains why survivors find it very difficult to recover unless they can find someone willing to listen to what happened to them. Kacandes notes that post-traumatic stress disorder (or PTSD) 'seems to require the creation of some kind of coherent narrative about the event or events that inflicted the trauma; this process is sometimes referred to as the translation of traumatic memory into narrative memory' (Kacandes 1999: 55).

Klein's explanation for this need lies in the process she described of working through repressed anxiety throughout life. People have not only the desire but the 'need' to tell their stories to others. Recognition in itself may be all that is required in these situations. This is the importance of *listening*, a key element in any relationship work. Brison talks about trauma narratives and the need that quickly becomes a dependency upon someone else to witness and acknowledge the story: 'Trauma undoes the self by breaking the ongoing narrative, severing the connections among remembered past, lived present, and anticipated future' (Brison 1999: 41).

It is through direct contact with the objects in the museum that this process of telling and listening can best be facilitated. A fascinating example of this took place when the Manchester Museum's way of working with refugees was adopted by a mental health team working with Afghani refugees in Peshawar, Pakistan (see p. 285 for this story). This all began five years ago when the Manchester Museum first worked with refugee women; the museum staff members involved found themselves listening in rapt silence as the women surprised all by proactively using the collections to tell their stories – through objects either brought from home or those produced from within the museum's collections.

Not knowing museum etiquette and touch prohibitions, the women disrupted precedent and took these things, in a real and metaphorical sense, into their own hands. They used the objects to explain what they knew about these collections and their relationship to their country of origin, subtly demanding recognition from the museum in terms of their knowledge of their own cultural heritage. Straightforward, one might say, but also important in terms of the reversal of authority. Even more interestingly, however, they used the objects as a means to convey to themselves, and to others, their individual feelings not only about what had happened to them in the recent and distant past, but what they imagined the future to become – how they might become in a future which they themselves were beginning to create in diaspora. As Bhabha put it, their 'presencing'.

What role is the object playing here? It is worth looking a little further into 'object relations' theory in psychoanalysis to attempt to answer this complex question. Melanie Klein made 'object-relations' central to the working through of emotional alienation and anxiety – key issues when working with diaspora communities in this country. Within this, Klein presents the role of the internal object and later, those who followed her, extended this to external objects. She emphasized the importance of:

- symbol formation
- fantasy
- play

If applied to museum collections, these are *uses* for the object that involve active agency on the part of the museum visitor when confronted with objects, as happened with the refugee women.

I would like to look briefly at this notion of working through internal and then external object relations in the work of Klein and then, most importantly, D.W. Winnicott, because I believe it can be seen as a crucial part of our potential relationship with people through objects in museums.

Internal object relations are, most importantly and formatively, with one's mother and other key relationships from early life onwards that become internalized, for good or ill, in the formation of the self. Klein described the processes at play as critical to the developing sense of self, and throughout life these relations with our internal 'objects' can be revisited and worked through, but are intrinsically ambivalent.

This ambivalence towards the internal, and later the external, object is key to what, in D.W. Winnicott's theory, would come to be seen as a relationship with the 'transitional' object. It was Winnicott (1971) who took up and further developed Klein's thesis into the realm of *external objects*, based on his own observations of children's play. Winnicott, who, upon carefully examining Klein's work with children, noted the importance of external objects in symbol formation and fantasy as children, and later adults too, continue throughout life to use objects to develop, review and redevelop a sense of self.

Jeanne Randolph developed Winnicott's theory of the transitional object and introduced for the first time the term 'the amenable object' (Randolph 1991). For Randolph, the amenable object is accessible for the viewer's needs and interpretation, offering an adaptive relationship. Thus, museum objects are amenable.

A 'Second Naiveté'

Within museums the stage is set for anything to happen – to experience things anew – and so acquire what Paul Ricoeur has termed a 'second naiveté' capable of conducting old inquiries in new ways (Kearney 1991: 5). Dr Ekpo Eyo, a Nigerian archaeologist, speaks of his personal experience of intermingling emotional and aesthetic response to objects: 'The more I looked at them, the more I described and handled them, the more emotionally attached I became' (cited in Vogel 1991).

We do not often know why it is that we are thus affected by particular objects, sometimes by something encountered in a museum that we may never have seen before. It doesn't make sense to us how this can be so. Bollas evocatively calls this,

> an intimate rendezvous with the 'unthought known' [in which] the shadow of the object falls upon the ego . . . the fundamental non-representational knowledge of being embraced by the . . . object. (Bollas 1987: 29–31)

In 2003 the Manchester Museum commissioned a film-maker, Kooj Chuhan, to develop a video project, involving a culturally diverse group of community members, entitled 'Rekindling Voices'. The individuals chose objects from within the display cases of the new anthropology gallery. These objects were then taken from their cases by conservation staff and the participants held 'conversations' with their selected objects, speaking 'to' rather than 'about' the objects. It was remarked that the individuals spoke to the objects as if to a part of themselves, an example of projective identification in action.[2]

In museums therefore, we may say that relations with objects involve a host of projections in the lifelong working through of anxiety, much of it in a creative form.

Manchester Museum's 'Rekindling Voices' video project[3] appears to provide evidence of this phenomenon in action in a form of projective identification. One participant, an Indian woman knows she is not the African jar, she respects the difference between herself and the jar, yet she is using the jar creatively, to work through her own feelings about exile and alienation. Objects elicit psychological reactions. People interact with things. They enter dialogues with the object. Objects become symbols or metaphors for specific activities, epochs, situations, relationships in people's real lives, and as such they 'recognize' objects in museums (Sandahl 1996). People, from the Somali women to this Indian women with the African jar, use objects to make relationality visible and manipulable. Hence, in a distinct split with the past, it is not the objects in museums, in and of themselves, that are important.[4] If we let them, objects in museums are 'amenable' to such use, suggesting a more in-depth use-value for the museum object. However, Winnicott stressed that it is not the object itself but the use to which it is put which determines its 'transitional' status. This idea of 'use' is therefore key to our approach to museum objects and to touch.

The Verb 'To Do'

Within the museum, individuals can utilize the object, analyse it, depict it, but most of all, they can *do* something with this object. They can even do something unexpected, as we saw in the 'Rekindling Voices' video project. As Randolph pointed out, the object is amenable. The museum can be an amenable space, a forum for such dialogues between the 'me' and 'not-me' to take place.

For this process to take place, however, the museum itself, as an institution, would have to become amenable. It is a challenge to didacticism within museum interpretation, and to systems of access and their current prohibitions within museums. Crucially, in a way that we can read in this context as a direct challenge to traditional museum practice, Winnicott says, 'To control what is outside one has to *do* things, not simply to think or to wish, and doing things takes time. Playing is doing' (Winnicott 1971: 41).

If we apply this to museums, simply suggesting this verb 'to do' as part of the museum experience immediately challenges the passive receptivity the museum experience engenders. For Winnicott, *play* became the key term: 'It is play that is universal: playing facilitates growth and therefore health ... in the service of communication with oneself and others' (ibid.: 41).

As phenomenology also emphasized, to experience something we have to *do* something with the things around us (Merleau-Ponty 1962). This often means that we have to pick up the object before us to 'experience it' fully. A demonstration of this powerful urge is well expressed by Canadian artist and writer Andrew Hunter, who describes a visit to the ornithology collections of the Manchester Museum in which he was researching a story on the now extinct bluebird that had been common to his part of Ontario:

> Even in their dry, slightly decayed, century-old state, they are beautiful. Not just the colours which are vivid and intense ... but also in their quiet repose ... .I want to touch them but don't, can't, because they are specimens, museum artefacts not to be engaged in this way ... When [the curator] leaves the room, I am tempted to pick one up, to hold it, gently touch its head and stroke its back. It has been killed and stuffed to be known but I'm not allowed to feel it. (Hunter 2004)

We want to learn about ourselves and the world in the way that is most natural to us but we are frustrated in the museum, prevented from picking up, taking hold of the desired object. For touch involves not only an exploration of surfaces but a total physical engagement with the object. To learn we must take the object into our hands and it is this transgressive act that crosses the boundaries into emotion.

Constructivist approaches to learning have long understood that passive touching is limited in that it can yield only a flow of stimuli which are difficult to identify. In order to learn we actively need to 'do' things with objects, as Winnicott reminded us;

to use what is available to us in exploring our world – to 'play', in much the same way as an artist does with available materials. This creative process might include, as Ybarra-Frausto puts it,

> [T]he utilization of available resources for syncretism, juxtaposition, and integration ... a sensibility attuned to mixtures and confluence ... a delight in texture and sensuous surfaces ... self-conscious manipulation of materials or iconography ... the combination of found material and satiric wit ... the manipulation of ... artefacts, code and sensibilities from both sides of the border. (Tomas Ybarra-Frausto, cited in Bhabha 1994: 7)

Artists in museums similarly disrupt precedent, and, for this reason, have much in common with diaspora communities, with children, and with all those for whom the touch taboo is unfamiliar – an obstacle to creativity and to learning and communication about the world and about the self in the world. It is *active touching* that is integral to the emotional, creative exploration of objects and therefore, of 'learning' itself. It involves, for the most part, the actual 'holding' of objects'.[5]

In active play, we 'do' things; we move, we gesture, using the object as a symbol in communicating with others. Vygotsky called this *symbolic play* a very complex system of 'speech' through gestures that communicate and indicate the meaning of playthings (Vygotsky 1978). From the infant's first experiences involving sensations that include sight, sound, kinaestheses, touch, smell, taste, the sensory relationship with objects, prior to language, return us to our earliest sensory experiences.

Sandahl wrote: 'Hairs and smells keep emerging after those that left or abandoned us – and brutally they activate memories' (Sandahl 1996). Objects thus invite a more integrative use of all senses, breaking down logocentrism and the visual dominance, which has become so prevalent in our culture. Ever since the refugee women left, the Manchester Museum has been working to make direct contact with objects through touch as central to its developing 'ways of working', and it has involved more than a few questions about our practices, prohibitions, prejudices – about ourselves.[6]

This work is not easy – it not only throws up many difficulties of a practical nature; but it also undermines the museum's culture of control, throwing up a challenge to the museum to let go, its gatekeepership of collections. If courageous, the museum can benefit from what the 'other', in the form of individuals from diaspora communities, has the potential to teach the museum, reminding us of our need in museums to take another look at our prejudices and prohibitions – as Spivak famously put it, begin the process of 'unlearning' them in order to allow ourselves to experience fully the world in partnership with our communities (Spivak 1995). It is happening more and more. The stranger, as Bhabha eloquently puts it, is now 'forcibly entering the house ... in order to invade, alarm ... to move beyond ... to turn the present into the 'past'; or ... to touch the future on its hither side' (Bhabha 1994: 16).

Making contact as fellow humans, 'touching' each other in the full emotional sense of the word – this is a corporeal invasion of the museum with ethical and political implications that burrows far beneath the surface of things. We live in a social world and each person is a 'relational construct'.

The phenomenologist Merleau-Ponty suggests that this is how we simultaneously experience myself and 'other', myself as 'other'. Levinas noted that 'instead of the distance between subject and object congenial to sight, touch restores the proximity of self and other ... it entails a more intimate relation with the world' (Levinas 1989: 79). It is for this reason that touch is so ethically important to museums. It is life as culture, as society, as the beginning of understanding, and in its symbolism and actuality, it begins the process of developing an ethical relationship with others. It is this experience that creates powerful relationships in museums such as that between diasporic people and museum workers coming together through making contact, firstly as people and then through joint experience of the handled, carefully passed-around personal or museum objects, warmed by touch. People in diaspora may come to the museum in search of clues, for something they can take hold of that says, like Khadra, one of the refugee women, when passing around her mother's dented metal pot with which she fled Somalia, 'Here I am, this is me'.

Notes

1. At the final gathering of the women for the culmination of the first 'Telling Our Lives' project at the Manchester Museum, one woman, a Nigerian, who was amongst those invited to attend, rebuked the Somali and Sudanese refugee women for hanging onto ideas of an idealized 'Africa' that their children may never know and to which the women themselves may never return. She added that, in fact, such an 'Africa' may never have existed.
2. Projective identification is a term first used by Melanie Klein in 1946. In her book *Psychoanalytical Diagnosis* (1994), Nancy Williams points out that projective identification combines elements of projection (attributing one's own feelings, thoughts and the motives of others).
3. The Manchester Museum's Rekindling Voices project involved the compilation of personal responses to objects on display. It was devised by members of the Museum's Community Advisory Panel working with local artist Kuljit Chuhan. This resulted in the Museum's Living Cultures gallery which features a screen-based interactive exhibit based on Rekindling Voices.
4. It is interesting to note that within the Portland Museum's North West Coast Indian consultation session with Tlingit tribal elders of which he wrote, Clifford mentions that 'stories and songs took centre stage', while the 'objects were left ... at the margin'. The objects were deemed to function as 'aides-memoire' (Clifford 1997: 189). Similarly, at the Manchester Museum, the Somali women's stories and songs took centre stage with the objects as their starting point.

5. See more on this in Winnicott (1971). Winnicott claims there is a direct development from transitional phenomena to playing and from this to cultural experiences.
6. The Manchester Museum has embedded the following sustainable 'ways of working' in its relationship work with communities, all involving direct access to collections:

    Out of the Museum: *Our City: Connections to Collections*
    Inside the Museum: Collective Conversations and Contact Zone
    Long-term partnerships: 'In Touch' Accredited Community Volunteering and The Community Advisory Panel

References

Bhabha H. (1991), 'The Other Question: Difference, Discrimination and the Discourse of Colonialism', in R. Ferguson et al. (eds), *Out There: Marginalisation and Contemporary Cultures*, New York: New Museum of Contemporary Art and London/Cambridge: MIT Press.

Bhabha, H.K. (1994), *The Location of Culture*, London/New York: Routledge.

Bollas, C. (1987), *The Shadow of the Object*, New York: Columbia University Press.

Brison, S.J. (1999),'Trauma narratives', in M. Bal, J. Crewe and L. Spitzer (eds), *Acts of Memory: Cultural Recall in the Present*, Hanover, NH: University Press of New England.

Chambers I. (1994), *Migrancy, Culture, Identity*, London: Routledge.

Clifford J. (1997), *Routes: Travel and Translation in the Late Twentieth Century*, Cambridge, MA: Harvard University Press.

Hunter A. (2004), *The Bluebird of Happiness*, unpublished story, March 2004.

Kacandes I. (1999), 'Narrative witnessing as memory work', in M. Bal, J.Crewe and L. Spitzer (eds), *Acts of Memory: Cultural Recall in the Present,* Hanover, NH/London: University Press of New England.

Kearney, Richard (1991), *Poetics of Imagining: From Husserl to Lyotard*, London: Routledge.

Klein, M. (1984[1932]), *The Psycho-Analysis of Children* (*The Writings of Melanie Klein*, Vol. 2), R. Money-Kyrle (ed.), A. Strachey (trans.), New York: Free Press.

Levinas E. (1989), 'Ethics as first philosophy', in S. Hand (ed.), *The Levinas Reader*, Oxford: Blackwell.

McWilliams, Nancy (1994), *Psychoanalytical Diagnosis: Understanding Personality Structure in the Clinical Process*, New York: Guilford Press.

Merleau Ponty, M. (1962), *Phenomenology of Perception*, London: Routledge.

Papastergiadis N. (1998), *Dialogues in the Diaspora, Essays and Conversations in Cultural Identity*, London: Rivers Oram.

Papastergiadis, N. (2000), *The Turbulence of Migration*, Cambridge: Polity Press.

Randolph, J. (1991), 'The Amenable Object', *Psychoanalysis and Synchronized Swimming and Other Writings on Art*, Toronto: YYZ Books.
Said E. (1990), 'Reflections on exile', in Russell Ferguson, Martha Gever, Trinh T. Minh-ha and Cornel West (eds), *Out There: Marginalisation and Contemporary Cultures*, Cambridge, MA: Harvard University Pres.
Sandahl, J. (1996). *Emotional Objects*, Vol. 6, Museum Ireland.
Sandler, J. and Sandler, A. (1978/1999), 'On the development of object relations and affects', *International Journal of Psychoanalysis* 59: 285–96.
Seth, V. (1994), *A Suitable Boy*, London: Phoenix.
Spivak, G.C. (1995), *The Spivak Reader*, D. Landry and G. MacLean (eds), London: Routledge.
Vogel S. (1991), 'Always true to the object', in I. Karp and S. D. Lavine (eds), *Exhibiting Cultures: The Poetics and Politics of Museum Display*, Washington, DC: Smithsonian Institute Press.
Vygotsky, L.S. (1978), *Mind and Society*, Cambridge, MA: Harvard University Press.
Winnicott, D.W. (1971), *Playing and Reality*, London: Tavistock Publications.

The Future

Touch and the Value of Object Handling: Final Conclusions for a New Sensory Museology

Devorah Romanek and *Bernadette Lynch*

Introduction

In this final chapter we will explore some of the key outcomes of the 'Touch and Object Handling' workshop series from which this volume originates.[1] The chapter will highlight cross-disciplinary themes, areas of best practice and recommend future directions of practice-led research.

> A mother and her child were having lunch in the beautiful garden of the Getty Museum in Los Angeles before going into the museum. They, like most people in the garden, were smelling and touching the flowers. The mother said 'It's wonderful to be able to touch all this, make the most of it now, before we go inside.'

This anecdote was recounted at the final 'Touch and the Value of Object Handling' conference by one of the participants who had overheard the conversation in the garden of the Getty Museum. The story was met with laughter from the conference participants, as it illustrates the challenge that many museums and their practitioners face, and which this project investigated – how to understand the value of touch.

Presentations were offered throughout the project that explored touch from a wide variety of disciplines, enabling participants to think and move across and beyond the boundaries that tend to restrict discipline-centred discourses. The 'Touch and the Value of Object Handling' end-of-project conference included the presentation of two final papers in the series, and two sets of four consecutively running discussion sessions.[2] The discussion sessions were facilitated by speakers who had given papers at previous workshops, by individual museum practitioners or by academics who had been part of developing the project. Although the conference marked the end of the project, it was not intended as an end point, but rather a further step towards understanding touch and the senses in general, and more specifically, the role of touch in museums. This summary of that conference offers an overview of the issues, illustrations of current best practice, and a crucial look towards what could come next. While all the workshops in this project consisted of the presentation of numerous papers, with the opportunity for questions and answers, the final conference offered an opportunity to generate much more discussion – to hear from the participants about what they were doing in their institutions, and to gather ideas towards future practice and research.

In all the sessions, it often seemed that the abstract, theoretical discussions that take place around the topic of touch in museums are at odds with the actual experience and practice that participants had to relate.[3] What became evident is that people experience objects *emotionally*. When speaking about their own experience of objects, whether academics, museum professionals or the general public, people frequently talk about how they *feel* as opposed to what they think, and the role of touch and *taking hold* so as to viscerally experience the emotional object appears to be of prime importance. What follows is an account of the discussions from the four sessions.

Session 1: Histories and Cultures of Touching

In order to understand how we have come to the discussion of touch and the value of object handling in museums, it is necessary to consider a history of sense hierarchies and aesthetic experience as related to museums. In this session, facilitated by Fiona Candlin (Birkbeck College, University of London and the British Museum), questions related to that history were raised, particularly through the lens of current practice. Discussion started with a look at how notions of touch in museums have shifted in recent history. It was noted that it was the passage of the UK Disability Discrimination Act (DDA) in 1995 that made this shift possible, enabling visitors to museums with disabilities to begin to engage with objects using senses besides sight, senses that were and frequently continue to be considered 'inferior' to sight. Before the passage of the DDA, the notion that touch may offer different ways of knowing that could be equally valued to the knowledge that sight offers, or to consider that touch and the other senses might inform vision, was not widely discussed. The passage of the DDA, coupled with shifts in anthropological discussions questioning the five-sense model of the senses with its privileging of the sense of sight, along with recent discoveries in neuropsychology related to the interconnectedness of all the senses, have begun to expand the thinking, discussion and practice related to this topic of sensory engagement with the material world in museums.

Changes in perspective regarding the senses have evolved to such an extent that in museums today, despite the traditional touch taboos, it is sometimes taken for granted, in theory if not in practice, that touch is a 'good thing'. Having arrived at this notion, the question was raised as to why this is so? Addressing this, participants often returned to the emotional experience that comes with engaging with an object through the sense of touch. When a person has the opportunity to handle an object, they can have the feeling that the object is a part of themselves or, conversely, that they are a part of the object – an experience of intimacy that would likely be denied were the object placed behind glass out of reach. One participant expressed this as being able to 'unlock' what is in the object. This metaphor – whereby feeling, knowledge or experience is 'locked' in an object – was a common invocation in all of the workshops. However, while people often focus on the positive emotions elicited through the process of being able to touch or hold an object, it was noted numerous times that the experience of handling an object can invite a range of emotions – from awe and wonder to dread and repulsion – and that keeping the full spectrum of emotional possibilities in mind is important in avoiding over-simplifying the potential experience of touch. The question thus raised was, in such a situation of permissible intimacy with an object, are people responding to being entrusted with handling an object (i.e. in the relationship between the authority figure and the person endowed with this trust), or are they responding to the object itself? This is one of the questions that participants in many of the sessions agreed needs further investigation. Assuming a satisfactory answer to this question, the question then becomes, does it actually matter what aspect of this experience people are responding to, particularly if we accept the importance of the 'relational' role of museums?

Further discussion segued into the political and sociological aspects of object handling, as it was observed that the issue of control as related to touch was implicit in all the discussions. One participant pointed out that when a museum puts an object on display, it is the museum who decides how visitors may view it, both literally and contextually. If a museum visitor is handed an object, that person can have the self-determining opportunity to decide how the object will be viewed, albeit within the context and constraints that the museum sets out for this engagement. This led to discussion of how, on an institutional level, access decisions are often very individual and idiosyncratic, having to do with the views of the head keeper, or the specific history of an institution, etc. These issues of power and control extend beyond the relationship of one institution and their visiting public, but also exist between institutions and collections, as more liberal touch access is afforded to some types of collections as opposed to others. One person suggested that, for example, natural history collections frequently offer more liberal access to touch than art collections. Access to whole collections often has an inverse relationship to its culturally defined and contextual value, as opposed to anything inherent in an object or collection of objects. Another participant offered an example of how initially curators at their institution, when solicited for objects for handling collections, gave objects that were only fragments. This of course, relates to the value placed on the fragments – fragments being perceived as less valuable than a whole object. Yet, fragments tend to be more useful or interesting to people who can conceptualize the whole (such as those with collections' expertise), and are therefore not overly useful in handling-collections. How we value objects can, of course, determine how much, if any, access is made available to collections. The issue of cultural value of objects as an inhibitor in terms of access to touch remains something worth further investigation.

In addition to culturally specific notions of value, there are, of course, a wide range of cultural notions of 'touch' and its importance. Participants discussed how many people consider objects in museums part of their spiritual or religious heritage or practice. Participants noted that when people encounter objects in museums, or indeed visit cultural heritage sites that they hold in high regard, they are often compelled, or feel a need to touch or kiss the object, to act out a ritual or leave an offering. It was also noted that this type of reverential touch is of course not always explicitly bound up with religion, as it can often be demonstrated in relation to all types of objects, whether such feelings be related to notions of the inherent value of an object or its associated history. An example of this was offered when one participant from a natural history museum related the way she observed numerous visitors, upon encountering an enormous Iguanodon dinosaur skeleton, reverently and almost reflexively touched the foot of the Iguanodon. The consideration of cultural norms in relation to touching objects spurred conversation onto various culturally specific taboos related to touch, and the complexities that this issue invited. The consideration of these culturally contextual types of touching or non-touching raised questions, such as: does this kind of touching have any bearing on the learning-through-touching sought through the assimilation of educational handling collections? Is this type of touch part of educational touching, or is it anathema to it? Although there was no consensus in answering these questions, participants did agree that this type of touching – or, in the case of honouring taboos, not-touching – is about establishing dialogues and therefore very important in the development of sustainable relationships between museums and their diverse audiences, based on trust.

Session 2: Touching Ghosts: Haptic Perception and Virtual Reality Technology

Could haptic interfaces and virtual objects ever serve as adequate substitutes for real touching?

This session, facilitated by David Prytherch (Birmingham Institute of Art and Design, University of Central England), decided the more fruitful approach to this question was to ask: how can this technology become more useful and offer different ways of engaging with objects and the material world? By avoiding establishing a hierarchy of the 'real' and the 'not-real', this session was able to investigate the 'what-if' in terms of the 'possible', and focus on the complementarity of this technology. Another key point that this session engaged with was the recognition that people now inhabit a multimedia world, with all the expectations that this brings, and that museums need to become familiar with the languages of these technologies in order to stay relevant. However, it was noted that these technologies can distract from what museums have to offer in terms of the authentic experiencing of collections, leading many participants to express that these technologies need to be employed with caution so that they don't only become a way of denying direct access to collections, and merely serving conservative, status quo agendas in museums.

Discussion began with considering the difference between the experience of holding an actual 3,000-year-old Roman glass vase and experiencing such a vase that has been created virtually. It was stated that the virtual haptic model may look the same, it may even feel vaguely the same, but the experience of it won't be the same; it won't have the same 'resonance'. However, as participants observed, the virtual version could enable tactile experiences that otherwise may not be possible, or permissible: the vase could be dropped, one could experience it shatter; the vase could be manipulated to create new textures, new combinations of properties. Perhaps it would be possible to create or invent a whole new aesthetic. Haptic technology could also make possible the modelling of bodily movement in relation to material culture – the passing on of embodied knowledge; someone without the skills or experience to create such a glass vase could enter into the virtual experience of such an activity based on somebody else's familiarity with a material or technique; one person could be working in a familiar way with a material, and somebody else could simultaneously feel what it is like to move around with that material in a virtual but simultaneous re-enactment of that experience. It was observed, however, that this type of body-modelling experience could also be inhibitive or proscriptive, something of which to be aware.

Elaborating on the concepts listed above, participants spoke of both their experiences with, and their hopes for, these technologies. Some of the ideas exchanged included using haptic technologies to create large databases of images, something like a design pattern book, with the attendant haptic qualities of the images enabling access to the texture and patterns of these material surfaces; this, in turn, would make possible an entirely different kind of search engine, leading to a new kind of research. A further application of this technology, already in use, entailed the creation of a virtual exhibition space that allowed visitors to rearrange objects in an exhibition – something of a power-sharing exercise in addition to a tactile one. Leading from this last point, it was discussed how the uses of these technologies are also about what we will never get to touch, what cannot be displayed – elaborations on imagining the potential of virtual collections of objects. Not only could previously impossible combinations of objects be arranged together, but they could also be experienced in locations in which it would not formerly have been possible. Even as it was considered that haptic technologies cannot replace the social experience of communal engagement and learning through objects, the examples participants offered also indicated new opportunities for virtual communal experiences and creative co-productions.

In talking about what directions or developments haptic technologies should take there was discussion about the dialectic nature of these technologies; the way that, as technology develops based upon what we know about ourselves, we in turn learn more about ourselves from these technologies. However, the point was made that a lot of what we experience sensually comes from what we as individuals know, consciously or unconsciously, and what is in one person's head is, of course, different to what is in another's. This makes it difficult to determine which of these developing technologies would be most useful as the answer to that question will necessarily be entirely individual. On the other hand, it was noted that the way our perception seems to work is that when you add one small sensory experience to another, the resulting sense of reality, of 'presence', increases exponentially, the sum being greater than the parts – or, as Prytherch put it: 'we embroider this wonderful rich reality out of these small bits of stuff', and these technologies can be an aide with bringing together more 'small bits of stuff' as we assemble an ever more elaborate sensory landscape.

Session 3: Touch Enhanced Learning for Under-represented Audiences

This session, facilitated by Jane Samuels and Laura Phillips (British Museum), addressed a wide range of questions, covering areas such as the development and structuring of touch and handling programmes, the nature and importance of 'authenticity' in this context and when and how it might be appropriate or constructive to use an original versus a replica with audiences, or vice versa. Additionally, the broader aims of enabling access through touch were addressed, and whether it needed always to be about 'learning' or whether 'enjoyment' was a legitimate aim in and of itself.

Participants agreed that in general, touch collections are created with specific audiences in mind in an effort to extend access to target groups, those who might normally be excluded from experiencing or participating in the museum. However, it was noted that the benefits to the museum are often ignored by supposing that those audiences with which the museum works in these programmes are the sole beneficiaries of this action, not recognizing that this can be a mutually beneficial experience, with the museum itself learning a great deal from these encounters. It was noted that the groups and individuals participating in inreach or outreach 'touch programmes' can challenge the museum to think in new ways. It was also observed that particular audiences may have specific needs, and therefore a range of objects and strategies of presentation and interpretation may be necessary to employ a successful 'touch programme'.

It was agreed that individual objects can offer very different types of experiences and opportunities; for example, some objects were more narrative by nature, while others more figurative, and so on. Participants did express varying viewpoints related to interpretation as the point was made that, without any contextualization, just being able to touch an object did not necessarily offer an understanding of, or meaningful engagement with, that object; conversely some participants bemoaned the over-contextualizing of objects, which can alienate and de-motivate visitors. This part of the discussion led some participants to express the need for an overarching philosophy or theoretical approach regarding touch in museums, while others suggested that leaving things open to investigation, whereby 'anything can happen', was a better approach. In fact, the unpredictability of responses to touch, and the fact that the museum cannot control the outcome, was seen as an advantage by some.

From here the discussion turned to the issue of the importance of being able to touch and handle original or 'authentic' objects, even as it was generally understood that notions of 'authenticity' are contentious and variable. It was also recognized that this was tied into further notions of power and entitlement, among other things. While it was cited here, as in other sessions, that handling an original object often allows for a much more intimate, powerful and emotional experience, it was pointed out by a participant that this is not always the case. He cited his experience as a blind person of being invited to touch original objects or things, particularly elements of historic buildings, archaeological sites, castles, etc., where the stone is often eroded and the carvings are indistinct. As he said of this experience: 'you know when you've felt one granite slab, you've felt them all really'. He related his experience in being able to handle restoration or replica objects or things, where the carvings are much more distinct, and how this experience can be much more coherent and emotionally powerful. He went on to elaborate that providing this type of touch experience to a *group* of people can also be very powerful:

> when you have a group of people with a shared commonality – a group of visually impaired people, for example – the experience of touch in groups can bring groups of people to a communal shared point of critical understanding, which is something that people who are sighted might take for granted. People who are sighted might stand in front of something and have a shared experience, a shared understanding by seeing what is in front of them. A group of visually impaired people may not have this experience if they are not touching.

Addressing the issue of the original versus the replica, one participant challenged the fetishization of the object in museums, and expressed her thoughts that museums were as much about critical engagement with an object as emotional connection. She suggested that museums should be about creating opportunities to unravel issues around an object intellectually – its production, its function, its intention – not about experiencing an object as a kind of icon. She reminded the group that sometimes it's an intellectual issue that one is trying to unravel about an object. Another participant also added that it isn't always practically possible to have the original, such as when, in a session, she deliberately banged a reproduction of a ship's biscuit very hard on a table in order to make a point about the nature of that object, something that would not be possible to do if using an original.

Focusing on the emerging conference tension between whether touching is about education and gaining knowledge or whether it is about enjoyment or emotion generated a lot of discussion, which was also echoed in other discussion groups. As one participant observed, if a museum offers objects to handle, and there is the risk they will be damaged, it is not enough to say 'well, people enjoy it'; at least it is not supposed to be a good enough reason, although many participants challenged this idea. It was expressed that failing to recognize 'enjoyment' as a legitimate aim in itself was related to a traditional suspicion and undervaluing of the 'base senses' and sensuality. There is a dominant view, that can be perpetuated by museums and one more recently critiqued by feminist theorists, which says that touch is about sensuality and more related to the 'female' and therefore dangerous, while sight, on the other hand, is seen as logical, rational and therefore male and 'in control'. Seen in this context, touch in museums also becomes a political act.

Session 4: Object Handling – Assessing Impact

As practitioners in the field, participants in this session, facilitated by Helen Chatterjee (University College London) and Devorah Romanek (University College London and the British Museum), were in strong agreement that 'touch' is unquestionably a worthwhile and beneficial exercise, and they could speak volumes about assessing this qualitatively and anecdotally. The notion that it is only through touch that certain and particular experiences are to be had or knowledge gained was a view validated by all. At the same time, participants agreed that expressing that in a quantitative way, such that it can be assessed, or even finding the words to fully illuminate or express what exactly transpires in object handling, can be a challenge. It was for this reason that the word 'impact' was selected as part of the title for the session, because 'impact' is a broader term than 'learning', and thus allows for a broader range of 'valuable' experiences. It was suggested that when people engage with objects they are not always 'learning' something in the strictest sense, but even if the impact is not about learning, the exchange that happens is a valuable one. However, this concept of 'value' within the touch experience still remains illusive. Alternatively, the concept of 'learning' needs to be broadened to include, for example, 'learning about oneself' through touch, as the museum engages in the use of collections in 'identity work'.

The question was asked, what happens when someone interacts with an object, and they feel disturbed by that encounter, but their curiosity is piqued at the same time? How does one assess the value of that type of engagement; how does one measure the worth or degree of someone's curiosity, or the complicated emotions that can be elicited? It was suggested that museums should be trying to get away from predetermining what will work and what sort of responses will be elicited from engagements with objects, that there should be less concern about measurable outcomes but, rather, objects should be put in play to see what happens. So long as people do respond and have a range of responses, should we be interested in predicting those responses? As Lynch pointed out, predicting responses could potentially be seen as unethical, yet the reality is that museums are compelled to evaluate these responses for various reasons – most particularly as part of the exercise of obtaining funding.[4]

There was agreement that it was necessary to find a way to assess usefully the value of touch as related to engaging with objects in museums, and there was discussion on a range of possible methods. One participant recounted the experience of his institution: they mailed out questionnaires about learning preferences and styles to visiting families before they came to the museum in order that, with the consent of the families, the museum could track them when they visited. This enabled the museum to evaluate the behaviour of the family as the visit unfolded, with the advantage of prior information as to the type of learner to expect from each member of the family. This preparation allowed the museum to evaluate specifically what was happening in interactions related to object handling. The museum found that when children in the family had very kinaesthetic learning styles they were able to lead their parents around interactives. This was so even though the parents may not have shared a kinaesthetic learning style (or have forgotten it). It was also found in the post-activity questionnaire that different type of learners derived different benefits from handling activities. Another participant described how, in her museum, it was found that when children engaged in object handling the museum facilitators were able to elicit in conversation with the children that they could differentiate between materials even when they were unable to identify the objects. The museum concluded that the children were clearly learning something through figuring out how to identify materials, but the museum was having trouble quantifying exactly *what* the children were learning. More importantly, does it matter if we can't precisely predict and subsequently identify the learning outcome? This example gave numerous participants the opportunity to share the concerns and difficulties they have with the Generic Learning Outcomes (GLOs) (see note 4) and the new Generic Social Outcomes (GSOs), leading one participant to reflect 'do you walk out of a play and get interrogated: "did you like the play, did you get something out of it, was it what you expected, what did you learn?"'

Regarding methodology, one participant observed that since object handling is an interactive, dynamic process that has to do with imagination, and with touch as humans in the world and with our bodies, traditional methodologies may not really be suitable for observation and analysis of this process. It was suggested that such a complex and ephemeral activity requires an equally complex and ephemeral approach to its evaluation, such as the new methodologies employed in visual sociology and visual ethnography – the dynamic process of recording and observing how the body engages in these activities, how people are moving or drawing, reading body language, and so forth, requiring a different type of evaluating than ticking boxes.

One participant recounted her museum's use of visual anthropology when testing a group of 6-year-old children: rather than using a questionnaire, the children reconstructed from felt the objects they had earlier handled, the results clearly indicating how much the children had learned and how efficacious the handling session had been. Someone else observed that this kind of re-creation after handling an object was almost like imagining one's own learning; furthermore, this imagining-your-own-learning is durational, which is another key aspect of the impact of object handling. Object handling has a long-lasting effect and relationship with memory, more so than text-based learning often has. It was further observed that object handling can also be a great leveller. One participant noted, while working with groups of children, that there are some children who are less verbally gifted than others, but who may have a more advanced ability to engage with the object in a tactile way.

This inspired a discussion of assessing impact in the use of authentic objects versus replicas, a discussion which echoed many points made in other sessions. Many participants thought that people had more 'invested' and were more focused when handling original objects, because they were more emotionally connected to the object, and therefore more present. It was also mentioned that, perhaps particularly when working with children, objects that have signs of handling often draw the children's attention. If an object is cracked, if it's damaged, if it's stained, if it shows the finger or thumbprints of the maker (as in some ancient clay objects), the object has a resonant history, and children can become connected to this person, or inspired by the object's story. Signs of wear and everyday use can capture the imagination of those of any age.

This led to discussion on the final point which was that object handling is not just about the object or the handler, but equally about the facilitator of such activities, as different people with different professional backgrounds bring different skills to such interactions. Many participants had experiences to share on this topic, recounting sessions that had been facilitated at various institutions, from museums to hospitals, and within each institution different people with different qualifications and levels of experience had facilitated sessions. The impact of this range of uses of touch, from learning and enjoyment to the therapeutic, were clearly and powerfully revealed, if only as yet anecdotally, but they all demonstrate the importance of a sensitive and skilled facilitator.

Conclusion: Touch, Exchange and Reciprocity

There is more to touch than meets the eye! As this workshop series demonstrated, the range of possibilities for the use of touch in museums is only just beginning to emerge. Touch is related to experiencing the world in its immediacy, not something we usually associate with museums. Touch therefore opens up the museum as an institution, to begin to engage more fully with the lived experience in the world around it, while delivering a critically important social function.

In March 2002, the Manchester Museum was approached by a University of Manchester research fellow and practising psychiatrist who was working on a project with refugees for the World Health Organization in Pakistan. He had read about the museum's 'Telling Our Lives' refugee women's project and was interested in the fact that, by then, the project had grown to include Afghani women who were recently arrived asylum seekers. His work in what was then the world's largest refugee camp in Peshawar, Pakistan, mainly concentrated on mental health issues for recently displaced Afghani women who were suffering from post-traumatic mental depression.

The doctor studied the methodology of the Manchester Museum's project, specifically in the use of object handling in the sharing of narrative experiences, and went on to try a similar method with the women in the camp, using an abandoned hut as a temporary 'museum' space. The refugee women in Peshawar were encouraged by mental health nursing volunteers assisting this doctor to bring their few possessions to 'storytelling' sessions, where they were asked to hand them round. These were objects of importance to them, those with which they had fled on their journey out of Afghanistan. The stories they began to tell were often painful but ultimately, it seems, of great assistance in the further care of these traumatized women. Evidently, the women also gained from the social aspect of these sessions, which were subsequently repeated.

That this hut in the centre of the refugee camp should be seen as a 'museum', pared down to its essentials, suggests a simple relationship with and communication through the handled object – a use for collections that may provide a good example for our complex cultural institutions in the West. In this way, museums may become useful 'social arenas', to borrow from Appadurai (1988), in which these transactions can take place, bringing together people from quite different cultural systems who share only the most minimal of understandings.

It is possible for museums to use touch to develop a relationship with diverse communities, based on a concept of 'mutuality' or reciprocity. Reciprocity in this context involves giving and receiving. Like the North West Coast American Indian *potlatch* (Mauss 1967), which is a system for the 'exchange of gifts', a festival where goods and services of all kinds are exchanged, hand to hand. The gifts exchanged in a potlatch are part of an elaborate reciprocal arrangement that conveys honour and respect on both parties. The museum could similarly and respectfully begin to serve this function, all the while learning from the exchange. As Sandahl recently suggested,

> What about the creation of a new museum machine, a facilitator of cross-cultural exchange to include respect and recognition – one that would involve hybridity, dialogism, transactionalism, always in motion, away from a controlling centre.[5]

Within this new museum, as this series of workshops on touch has shown, we may begin truly to accommodate the complexity of sensory experience that can be found in the interaction of bodies and things in the world outside the museum's doors.

Notes

1. 'Touch and the Value of Object Handling in Museums' workshop series held November 2006 to May 2007, organized by UCL Museums & Collections, funded by the Arts and Humanities Research Council. (http://www.museums.ucl.ac.uk/learning/touch_research/)
2. In keeping with the inter-disciplinarity of the series, and in order to ground the final discussions, these two papers were presented from very different disciplines – one by Andrew Wimperis (University of Birmingham) from the discipline of neuropsychology, and one by Bernadette Lynch from the discipline of anthropology and museology.
3. The four discussion sessions began with facilitators presenting three or four prompt questions to get the discussion in motion, but the discussions often ventured into interesting territory more illuminating than these prompt questions could solicit or imply.
4. The Museum Libraries and Archives (MLA) Council have developed the Generic Learning Outcomes (GLOs) as part of the Inspiring Learning for All initiative, to try to include (and record) these intangible, affective responses to the museum experience. See the MLA website (www.mla.gov.uk) for information on the GLOs.
5. Jette Sandahl, speaking at the 'Collections, Cultures, Change' conference, Centre for Museology, University of Manchester, November 2003.

References

Appadurai, Arjun (1994), 'Commodities and the politics of value', in Susan M. Pearce (ed.), *Interpreting Objects and Collections*, London: Routledge, pp. 76–91.

Mauss, Marcel (1967 [1923]), *The Gift: Forms and Functions of Exchange in Archaic Societies*, New York: Norton.

Index

Note: italicized page numbers indicate illustrations

Lightning Source UK Ltd.
Milton Keynes UK
UKOW06f0014070316

269731UK00009B/361/P

9 781847 882387